Stochastic Analysis in Mathematical Physics

Stochastic Analysis in Mathematical Physics

Proceedings of a Satellite Conference of ICM 2006

Lisbon, Portugal   4 – 8 September 2006

Editors

Gerard Ben Arous
Courant Institute, New York University, USA

Ana Bela Cruzeiro
GFMUL & IST (TUL), Portugal

Yves Le Jan
Universitè Paris 11, France

Jean-Claude Zambrini
GFMUL, University of Lisbon, Portugal

NEW JERSEY · LONDON · SINGAPORE · BEIJING · SHANGHAI · HONG KONG · TAIPEI · CHENNAI

Published by
World Scientific Publishing Co. Pte. Ltd.
5 Toh Tuck Link, Singapore 596224
USA office: 27 Warren Street, Suite 401-402, Hackensack, NJ 07601
UK office: 57 Shelton Street, Covent Garden, London WC2H 9HE

British Library Cataloguing-in-Publication Data
A catalogue record for this book is available from the British Library.

STOCHASTIC ANALYSIS IN MATHEMATICAL PHYSICS
Proceedings of the Satellite Conference of the ICM 2006

Copyright © 2008 by World Scientific Publishing Co. Pte. Ltd.

All rights reserved. This book, or parts thereof, may not be reproduced in any form or by any means, electronic or mechanical, including photocopying, recording or any information storage and retrieval system now known or to be invented, without written permission from the Publisher.

For photocopying of material in this volume, please pay a copying fee through the Copyright Clearance Center, Inc., 222 Rosewood Drive, Danvers, MA 01923, USA. In this case permission to photocopy is not required from the publisher.

ISBN-13 978-981-279-154-2
ISBN-10 981-279-154-X

Printed in Singapore.

PREFACE

In the Summer of 2006, an International Conference, satellite of the International Congress of Mathematicians (ICM2006, Madrid), took place in Lisbon and was organized by the Group of Mathematical Physics of the University of Lisbon (GFMUL).

We decided to entitle "Stochastic Analysis in Mathematical Physics" the elaboration of the ideas presented there by some of the participants.

The last ten years were witness to a remarkable penetration of the methods of Stochastic Analysis in all fields of Mathematical Physics. What was regarded by many, not so long ago, as a set of esoteric tools, turned into a fundamental component of our understanding of natural phenomena. The works collected here illustrate the versatility of those stochastic methods and we warmly thank the authors.

The Group of Mathematical Physics is sponsored by the Portuguese Foundation for Science and Technology, specifically, for this conference, by Project POCTI/0208/2003, co-financed by FCT/OE and FEDER, as well as Project POCI/MAT/55977/2004.

We thank also the Foundation Calouste Gulbenkian and the FLAD (Luso-American Foundation) for their support.

The Editors,

G. Ben Arous (*New York*)
A. B. Cruzeiro (*Lisbon*)
Y. Le Jan (*Paris*)
J. C. Zambrini (*Lisbon*)

October 11, 2007

LIST OF PARTICIPANTS

Gérard Ben Arous	École Polytechnique Fédérale de Lausanne, Switzerland
Jean Bertoin	Université Paris VI, France
Erwin Bolthausen	Universität Zürich, Switzerland
Fernanda Cipriano	GFMUL/Univ. Nova de Lisboa, Portugal
Ana Bela Cruzeiro	GFMUL/Instituto Superior Técnico, Portugal
Krzysztof Gawedzki	École Normale Supérieure de Lyon, France
Patrícia Gonçalves	IMPA, Brazil
Maria Gordina	University of Connecticut, USA
Vadim Iourinski	Universidade da Beira Interior, Portugal
Antii Kupiainen	University of Helsinki, Finland
Yves Le Jan	Université Paris-Sud 11, France
Thierry Lévy	École Normale Supérieure, France
Wu Liming	Université Blaise Pascal-Clermont II, France
Piotr Lugiewicz	Imperial College London, UK
Paul Malliavin	Université Paris VI, France
Jonathan Mattingly	Duke University, USA
Stanislav Molchanov	University of Northern Carolina – Charlotte, USA
Ana Filipa Pontes	ISEP, Portugal
Rolando Rebolledo	Pontificia Universidad Católica, Chile
Jorge Rezende	GFMUL/Universidade de Lisboa, Portugal
Christophe Sabot	École Normale Supérieure de Lyon, France
Ambar Sengupta	Louisiana State University, USA
Alain-Sol Sznitman	ETH-Zürich, Switzerland
Eric Vanden-Eijnden	Courant Institute of Math. Sciences, USA
Jean-Claude Zambrini	GFMUL/Universidade de Lisboa, Portugal

CONTENTS

Preface   v

Participants   vii

Stochastic parallel transport on the d-dimensional torus   1
   Ana Bela Cruzeiro and Paul Malliavin

Riemannian geometry of $\mathrm{Diff}(S^1)/S^1$ revisited   19
   Maria Gordina

Ergodic theory of SDE's with degenerate noise   30
   Antti Kupiainen

Dynkin's isomorphism without symmetry   43
   Yves Le Jan

Large deviations for the two-dimensional Yang-Mills measure   54
   Thierry Lévy

Laplace operator in networks of thin fibers: Spectrum near the threshold   69
   S. Molchanov and B. Vainberg

Adiabatic limits and quantum decoherence   94
   Rolando Rebolledo and Dominique Spehner

Gauge theory in two dimensions: Topological, geometric and probabilistic aspects   109
   Ambar N. Sengupta

Near extinction of solution caused by strong absorption on a
fine-grained set 130
   V. V. Yurinsky and A. L. Piatnitski

1

Stochastic parallel transport on the d-dimensional torus

Ana Bela Cruzeiro

Departamento de Matemática, Instituto Superior Técnico,
Universidade Técnica de Lisboa,
Av. Rovisco Pais, 1049-001 Lisboa, Portugal
and
Grupo de Física-Matemática, Universidade de Lisboa
Av. Prof. Gama Pinto, 2, 1649-003 Lisboa, Portugal
E-mail: abcruz@math.ist.utl.pt

Paul Malliavin

10 Rue S. Louis en-l'Ile, 75004 Paris, France

In Ref. 2 V. Arnold has shown that the Euler flow can be identified with a geodesic on the group G of volume preserving diffeomorphisms with respect to the L^2 metric. Following this approach, the geometry of G plays a fundamental role in hydrodynamics and is important for instance in the study of the stability of the fluids motion. It has been developed by many authors, one of the first being Ref. 5, see also Ref. 3 and references therein.

In this paper we consider the d-dimensional torus as the underlying manifold, we compute the constants of structure and the Christoffel symbols of the corresponding group of volume preserving L^2 diffeomorphisms.

Using infinite dimensional stochastic analysis we construct the stochastic parallel transport on G along Brownian paths where some weights on the Fourier modes are considered. Then there is a matrix which describes the energy transfer between modes: its exact computation in the two dimensional case has been done in Ref. 4, where a machinery already developed in Ref. 1 in the context of the Virasoro algebra was used. Here, for the d-dimensional case, we prove the existence of such matrix and establish some qualitative estimates.

Keywords: Stochastic parallel transport, geometry of diffeomorphisms group

1. Basis of Lie algebra of vector fields with vanishing divergence

We shall denote by G the group of measure preserving diffeomeophisms of the torus T^d and by $\mathcal{G}$ its Lie algebra, consisting of real vector fields on T^d with vanishing divergence.

We look for an L^2-orthonormal basis of *complex* vector fields with vanishing divergence. We use the key fact of the invariance of this space by the translation operator:
$$[(\tau_\varphi)_* Y](\theta) := Y(\theta - \varphi).$$
Define
$$Y^n := \frac{1}{\pi^d} \int_{T^d} [(\tau_\varphi)_* Y] \exp(in.\varphi) \, d\varphi, \quad n = (n_1, \ldots, n_d) \in Z^d.$$
Then Y^n has a vanishing divergence; furthermore
$$((\tau_{\varphi_0})_* Y^n)(\theta) = \frac{1}{\pi^d} \int_{T^d} Y(\theta - \varphi - \varphi_0) \exp(in.\varphi) \, d\varphi.$$
Make the change of variable $\varphi' - \varphi_0 = \varphi$; then
$$((\tau_{\varphi_0})_* Y^n)(\theta) = \exp(-in.\varphi_0) Y^n)(\theta).$$
Considering this identity at $\theta = 0$ we find $Y^n(-\varphi_0) = \exp(-in.\varphi_0) Y^n(0)$. Therefore we can look for a basis of vector fields with vanishing divergence of the form
$$z_n \exp(in.\theta), \quad z_n \text{ is a fixed vector in } R^d.$$
As we have $\operatorname{div}(z_n \exp(in.\theta)) = i\, n.z_n \, \exp(in.\theta)$, we get

Proposition 1.1. *The vector fields of the form*
$$z_k \exp(i\, k.\theta), \quad k.z_k = 0 \tag{1.1$_a$}$$
generate the vector space of complex vector fields with zero divergence, and vector fields asssociated to two distinct values of k are orthogonal.

To find an orthonormal basis we have to find for each k an orthonormal basis of the space $V_k := \{z \in C^d : k.z = 0\}$; define
$$\mathcal{E}_k := \{x \in R^d : k.x = 0\}, \text{ then } \mathcal{E}_{-k} = \mathcal{E}_k; \tag{1.1$_b$}$$
and $V_k = \mathcal{E}_k \otimes C$. We obtain the wanted orthonormal basis by picking, for $\forall k \neq 0$, an orthonormal basis $\epsilon_k^1, \ldots, \epsilon_k^{d-1}$ of each $\mathcal{E}_k$; we make the convention to take
$$\epsilon_{-k}^\alpha := \epsilon_k^\alpha. \tag{1.1$_c$}$$

Proposition 1.2. *Denote $\tilde{Z}^d$ a subset of Z^d such that each equivalence class of the equivalence relation on Z^d defined by $k \simeq k'$ if $k + k' = 0$ has a* unique repesentative in $\tilde{Z}^d$; denote $\mathcal{H}_r$ the Hilbert space of square integrable real vector fields on T^d with a vanishing divergence, then

$$\left\{ \epsilon_k^\alpha \cos k.\theta, \quad \epsilon_k^\alpha \sin k.\theta \right\}_{k \in \tilde{Z}^d,\ \alpha \in [1,d-1]} \quad (1.2)$$

constitute an orthonormal basis of $\mathcal{H}_r$.

Proof. We have

$$Y = \sum_k z_k \times \exp(i\ k.\theta), \quad z_k \in \mathcal{E}_k, \text{ with the reality condition } z_{-k} = \bar{z}_k$$
$$(1.3_a)$$

and $\|Y\|_{L^2}^2 = \sum_k \|z_k\|_{R^d}^2$.

We can write the complex series (1.3_a) as a real series by grouping terms:

$$Y = \sum_{k \in \tilde{Z}^d} \left(z_k \times \exp(ik.\theta) + \bar{z}_k \times \exp(-ik.\theta) \right), \quad z_0 \in R^d,\ z_k \in \mathcal{E}_k \otimes C. \quad (1.3_b)$$

For $k \in \tilde{Z}^d$, writing $z_k = \alpha_k - i\ \beta_k$, $\alpha_k, \beta_k \in \mathcal{E}_k$, we get

$$Y = \sum_{k \in \tilde{Z}^d} \left(\alpha_k \times \cos(ik.\theta) + \beta_k \times \sin(ik.\theta) \right), \quad \alpha_0 \in R^d,\ \alpha_k, \beta_k \in \mathcal{E}_k. \quad (1.3_c)$$

We want to write automatically the passage from the representation (1.3_a) to the representation (1.3_c). We shall formalize the following elementary Euler identities

$$\cos k.\theta = \frac{1}{2}\left(\exp(ik.\theta) + \exp(-ik.\theta) \right), \quad \sin k.\theta = \frac{1}{2i}\left(\exp(ik.\theta) - \exp(-ik.\theta) \right).$$
$$(1.3_d)$$

Consider the group ς generated by the symmetry $\tau : k \mapsto -k$ on Z^d; then ς is a group of order 2. Denote χ the character on ς which takes the value -1 on τ and the value 1 on the identity. The powers of χ generate a group of order two; then there exists a coupling of duality $\varsigma \times \hat{\varsigma} \mapsto \{1,-1\}$, coupling denoted $< *, * >$, where $\hat{\varsigma}$ is the dual group of ς. Given a function a defined on Z^d, its ς-*Fourier transform* is defined as

$$\hat{\sigma}a_k = \frac{1}{2}\sum_{\sigma \in \varsigma} a_{\sigma(k)} \times <\sigma, \hat{\sigma}>, \quad k \in \tilde{Z}^d.$$

Starting from (1.3_c) we define a function ψ on $\hat{\varsigma} \times \tilde{Z}^d$ by

$$\alpha_k =: \psi_k(e), \quad \beta_k =: \psi_k(\chi), \quad k \in \tilde{Z}^d,$$

then we get the universal formula describing the passage from (1.3_a) to (1.3_c)

$$\psi_k(\hat{\sigma}) = \frac{1}{\sqrt{<\tau,\gamma>}} \times {}^{\hat{\sigma}}z_k, \quad k \in \tilde{Z}^d. \qquad (1.3_e)$$

If we apply (1.3_e) to the function $\exp(ik.\theta)$ we find back the Euler identities (1.3_d). □

2. Constants of structure of the Lie algebra and Christofell symbols

The bracket of complex vector fields has the following expression:

$$\left[z_k \times \exp(i\,k.\theta)\,,\;z_s \times \exp(i\,s.\theta)\right] = \left((z_k.s)\,z_s - (z_s.k)\,z_k\right) \times \exp\left(i\,(k+s).\theta\right);$$

as $(z_k.s)(k+s).z_s - (z_s.k)(k+s).z_k = 0$, $\left((z_k.s)\,z_s - (z_s.k)\,z_k\right) \in \mathcal{E}_{k+s}$.
Let

$$b_{k,s}: (\mathcal{E}_k \otimes C) \times (\mathcal{E}_s \otimes C) \mapsto \mathcal{E}_{k+s} \otimes C \text{ be defined by}$$

$$b_{k,s}(z_k, z_s) = \left((z_k.s)\,z_s - (z_s.k)\,z_k\right); \text{ hence}$$

$$\left[z_k \times \exp(i\,k.\theta)\,,\;z_s \times \exp(i\,s.\theta)\right] = b_{k,s}(z_k, z_s) \times \exp(i\,(k+s).\theta). \quad (2.1_a)$$

Remark that, granted the identification (1.1_b), we have

$$b_{-k,-l} = -b_{k,l}. \qquad (2.1_b)$$

The function $b_{k,l}$ is defined on $\mathcal{E}_k \times \mathcal{E}_l$ and takes its values in $\mathcal{E}_{k+l}$. In the orthonormal basis ϵ_k^α, ϵ_l^β, ϵ_{k+s}^γ it is expressed as

$$[b_{k,l}]_{\alpha,\beta}^\gamma \text{ or more intrinsically } b_{k,l} \in (\mathcal{E}_k \otimes \mathcal{E}_l \otimes \mathcal{E}_{k+l}) \otimes C, \qquad (2.1_c)$$

this last indentification being possible granted the euclidean structure of the $\mathcal{E}_*$.

Let Y, Y' be two real vector fields

$$Y = \sum_{k \in \tilde{Z}^d} \left(z_k \times \exp(ik.\theta) + \bar{z}_k \times \exp(-ik.\theta)\right),$$

$$Y' = \sum_{s \in \tilde{Z}^d} \left(z'_s \times \exp(is.\theta) + \bar{z}'_s \times \exp(-is.\theta)\right),$$

$$[Y, Y'] = \sum_{k,s \in \tilde{Z}^d} A_{k,s} \text{ where}$$

$$A_{k,s} := \left[z_k \times \exp(i\, k.\theta) + \bar{z}_k \times \exp(-i\, k.\theta),\ z'_s \times \exp(i\, s.\theta) + \bar{z}'_s \times \exp(-i\, s.\theta) \right] \tag{2.1$_d$}$$

$$= b_{k,s}(z_k, z'_s) \times i \exp\big(i\ (k+s).\theta\big) + b_{-k,s}(\bar{z}_k, z'_s) \times i \exp\big(i\ (-k+s).\theta\big)$$

$$+ b_{k,-s}(z_k, \bar{z}'_s) \times i \exp\big(i\ (k-s).\theta\big) + b_{-k,-s}(\bar{z}_k, \bar{z}'_s) \times i \exp\big(-i\ (k+s).\theta\big)$$

Denote by δ the Kronecker symbol and define the *constants of structure of the complex Lie algebra* $\mathcal{H}$ by

$$c^r_{k,s} = b_{k,s}\, \delta^r_{k+s}. \tag{2.1$_e$}$$

Previously the component of a vector was depending only upon indices $k \in Z^d$. Now the constants of structure depend upon three indices; the natural group of symmetry is ς^3, group which has for dual group $\hat{\varsigma}^3$, the coupling being given by

$$<\sigma,\ \hat{\sigma}> = xs \prod_{i=1}^{3} <\sigma_i,\ \hat{\sigma}_i>$$

Define

$$\hat{\sigma} c^r_{k,s} = \frac{1}{8\sqrt{<\tau,\ \gamma>}} \sum_{\sigma \in \varsigma^3} c^{\sigma_3(r)}_{\sigma_1(k), \sigma_2(s)} <\sigma,\ \hat{\sigma}>, \tag{2.2}$$

$$\tau = (\tau_1, \tau_2, \tau_3),\quad k, s, r \in \tilde{Z}^d,\quad \hat{\sigma} \in \hat{\varsigma}^3.$$

Remark 2.1. We have $c^{\sigma_3(r)}_{\sigma_1(k),\sigma_2(s)} \in \mathcal{E}_{\sigma_1(k)} \otimes \mathcal{E}_{\sigma_2(s)} \otimes \mathcal{E}_{\sigma_3(r)} \otimes C$ which by (1.1$_b$) is equal to $\mathcal{E}_k \otimes \mathcal{E}_s \otimes \mathcal{E}_r \otimes C$; therefore all the elements of the sum (2.2) belong to the same vector space.

Theorem 2.1. *Define a function ψ giving the expression of the bracket in the basis*

$$x_k \cos k.\theta,\ y_k \sin k.\theta,\quad k \in \tilde{Z}^d,\quad x_k,\ y_k \in \mathcal{E}_k\ as$$

$\psi^r_{k,s}(1,1,1)$ *is the term which gives the contribution of the terms in* $\cos k.\theta,\ ,\cos s.\theta$ *on the component on* $\cos r\theta$; $\psi_{k,s}(1,1,\chi)$ *the component of the same term on* $\sin r\theta$ *and so on. Then we have the universal formula*

$$\psi^r_{k,s}(\hat{\sigma}) = \hat{\sigma} c^r_{k,s}. \tag{2.3}$$

Proof. We define an induction on the number of components. We have proved the universal formula when we have a single component. Assume that we have proved it for tensors with a number of indices $< p$. Consider a tensor with p indices. Fixing the last index we obtain a tensor with $(p-1)$ indices for which the universal formula holds true; fixing the first $(p-1)$ components we obtain a vector for which the universal formula holds true; finally the universal formula stays stable by cartesian product. □

Christofell symbols

The Christofell symbol in the complex basis are defined in tems of the structural constant

$$2\Gamma^l_{k,s} := c^l_{k,s} - c^k_{s,l} + c^s_{l,k} = b_{k,s}\delta^l_{k+s} - b_{s,l}\delta^k_{s+l} + b_{l,k}\delta^s_{k+l} \qquad (2.4_a)$$

Remark 2.2. If $kls \neq 0$ at most a single Kronecker symbol does not vanish and in the sum (2.4_a) at most one term is different from zero; for this reason we have not to worry about adding objects which belong to different vector spaces. To avoid to check constantly facts of this nature we shall use the canonical injections $j_k : \mathcal{E}_k \mapsto R^d$, which induce a canonical injection $j_k \otimes j_s \otimes j_l : \mathcal{E}_k \otimes \mathcal{E}_s \otimes \mathcal{E}_l \mapsto R^d \otimes R^d \otimes R^d$; all the considered objects will belong to the fixed vector space $R^d \otimes R^d \otimes R^d$.

We obtain the Christofell symbols in the real basis by applying the universal formula (2.2):

$$[\hat{\sigma}\Gamma]^l_{k,s} := \frac{1}{8\sqrt{<\tau, \gamma>}} \sum_{\sigma \in G^3} \Gamma^{\sigma_3(l)}_{\sigma_1(k),\sigma_2(s)} \times <\sigma, \hat{\sigma}>, \qquad (2.4_b)$$

$$<\sigma, \hat{\sigma}> := \prod_{i=1}^{3} <\sigma_i, \hat{\sigma}_i>,$$

where $k, s, l \in \tilde{Z}^d$.

Proposition 2.1. *For every $\gamma \in \hat{\sigma}^3$ we have $[\hat{\sigma}\Gamma]^l_{k,s} \in \mathcal{E}_k \otimes \mathcal{E}_s \otimes \mathcal{E}_l$.*

In the usual theory Christofell coefficients are scalar; to realize this situation we have to pick a basis $\epsilon^\alpha_k, \alpha \in [1, d-1]$ of each $\mathcal{E}_k$; then the coefficient $[^\gamma\Gamma]^l_{k,s}$ will give rise to $(d-1)^2$ scalar coefficients. We prefer to avoid this explicit computation by introducing the following vector valued tensor calculus. We consider the vector bundle $\mathcal{F}$ of basis $\tilde{Z}^d$ defined as

$$\mathcal{F} := \bigcup_{k \in \tilde{Z}^d} \mathcal{E}_k \text{ and its dual bundle defined as } \mathcal{F} := \bigcup_{k \in \tilde{Z}^d} \mathcal{E}^*_k;$$

of course the euclidean structure of $\mathcal{E}_k$ defines an identification of $(\mathcal{E}^*_k)^*$ with $\mathcal{E}_k$. In tensor calculus we shall forget this identification. A tensor q-times covariant and p-times contravariant is by definition a section of the vector bundle $[\mathcal{F}^*]^{\otimes q} \bigotimes [\mathcal{F}]^{\otimes p}$

Proposition 2.2. *Fix $\hat{\sigma} \in \hat{\varsigma}^3$, then we can consider $[\hat{\sigma}\Gamma]^*_{*,*}$ as a tensor 2-times covariant and 1-times contravariant.*

Proof. Use Prop. 2.1 and the identification between $[\mathcal{E}_k]^*$ and $\mathcal{E}_k$. □

Tensorial contraction

Given ξ^* a contravariant tensor an η_* a covariant tensor their contraction is

$$\sum_{k \in \tilde{Z}^d} \text{trace } \xi^k \otimes \eta_k := \sum_{k \in \tilde{Z}^d} <\xi^k, \eta_k>$$

We can contract two contravariant (resp. covariant) indices using the identification of $\mathcal{E}_k$ with $\mathcal{E}_k$ granted the underlying euclidean metric. We shall emphasize the use of this euclidean structure by the notation

$$\sum_{k \in \tilde{Z}^d} \text{trace }_{k,k} \xi^k \otimes \eta^k = \sum_{k \in \tilde{Z}^d} (\xi^k \mid \eta^k)_{\mathcal{E}_k}$$

Proposition 2.3. *Fix $k \in Z^d$ and fix $z \in \mathcal{E}_k$; denote $\Gamma(z)$ the corresponding operator on $\mathcal{F}$. Then $\Gamma(z)$ is antisymmetric operator.*

Proof. Choose z_*, z'_*, z''_* such that $z_{-q} = \bar{z}_q$, $z'_{-s} = \bar{z}'_q$, $z''_{-q} = \bar{z}_q$. Define the euclidean scalar product on $\mathcal{E}_* \otimes C$ by the algebraic prolongation of the euclidean scalar product on $\mathcal{E}_*$: we obtain a scalar product which is C-linear relatively to the second factor. Using formula (2.4_a), introduce the expression

$$\sum_{k,s} \bigl(b_{k,s}(z_k, z'_s) \mid z''_{k+s}\bigr) - \bigl(b_{s,k-s}(z'_s, z''_{k-s}) \mid z_k\bigr) + \bigl(b_{s-k,k}(z''_{s-k}, z_k) \mid z'_s\bigr).$$

(2.5_a)

Fix z and show the antisymmetry in z', z''; firstly

$$b_{s,k-s}(z'_s, z''_{k-s}) = -b_{k-s,s}(z''_{k-s}, z'_s)$$

according to the expression

$$b_{k,l}(z_k, z_s) = (z_k.s)\, z_s - (z_s.k)\, z_k$$

therefore the antisymmetry of the middle term of (2.5_a) is obtained.

In the third term of (2.5_a) making the change of index of summation $(s-k) \mapsto q \simeq s$, we obtain that the contribution of the first and third term of (2.5_a) can be written as

$$\sum_{k,s} \big(b_{k,s}(z_k, z'_s) \mid z''_{k+s}\big) + \big(b_{s,k}(z''_s, z_k) \mid z'_{k+s}\big).$$

On the other hand,

$$\big(b_{k,s}(z_k, z'_s) \mid z''_{k+s}\big) + \big(b_{s,k}(z''_s, z_k) \mid z'_{k+s}\big)$$
$$= (z_k.s)\,(z'_s \mid z''_{k+s}) - (z'_s.k)\,(z_k \mid z''_{k+s}) -$$
$$(z_k.s)\,(z''_s \mid z'_{k+s}) + (z''_s.k)\,(z_k \mid z'_{k+s})$$
$$= -(z'_s.k)\,(z_k \mid z''_{k+s}) + (z''_s.k)\,(z_k \mid z'_{k+s}),$$

expression which is obviously antisymmetric in z', z''. □

Corollary 2.1. *For fixed k the matrices*

$$^{\hat{\sigma}}\Gamma^*_{k,*} \text{ are antisymmetric } \forall \hat{\sigma} \in \hat{\varsigma}^3. \qquad (2.5_b)$$

Proof. As $^\gamma\Gamma^*_{k,*}$ is obtained by averaging an antisymmetric operator, it is antisymmetric. □

Notation 2.1. Given ξ^k a contravariant tensor, we denote the following antisymmetric operator

$$\sum_k {}^{\hat{\sigma}}\Gamma^*_{k,*}\,\xi^k =: {}^{\hat{\sigma}}\mathbf{\Gamma}(\xi)^*_* \ . \qquad (2.6)$$

Proposition 2.4. *The component of the Christofell symbol on the three cosines type vanishes.*

Proof. This component is equal to $^{\hat{\sigma}_0}\Gamma$, where $\hat{\sigma}_0 = (1,1,1)$; remark that $<\tau\sigma,\ \hat{\sigma}_0> = <\sigma,\ \hat{\sigma}_0>$ and that $\delta^{\tau(l)}_{\tau(k),\tau(s)} = \delta^l_{k,s}$. Therefore the sum of twenty four terms defining $^{\hat{\sigma}_0}\Gamma$ can be split in the sum of twelve terms each having in factor the following expression $\delta^{l'}_{k',s'} \times (b_{k',s'} + b_{\tau(k'),\tau(s')})$ and (2.1_b) finishes the proof. □

3. Stochastic parallel transport, symmetries of the noise

Consider for each $k \in \tilde{Z}^d$ the complex brownian motion $\zeta_k(t)$ associated to the natural hermitian metric on $\mathcal{E}_k \otimes C$; all these brownian motion are taken to be independent on the system of relations

$$\zeta_{-k}(t) = \bar{\zeta}_k(t), \text{or in real terms } \zeta_k = x_k + i\,y_k(t), \quad k \in \tilde{Z}^d\,, \qquad (3.1)$$

the x_k, y_k being independent brownian motion on $\mathcal{E}_k$. Define

$$\boldsymbol{\Gamma}(dx_k(t)) = \sum_{\gamma_1=1} {}^{\gamma}\boldsymbol{\Gamma}(dx_k(t)), \quad \boldsymbol{\Gamma}(dy_k(t)) = \sum_{\gamma_1=-1} {}^{\gamma}\boldsymbol{\Gamma}(dy_k(t)), \quad (3.2_a)$$

where the sumation is taken on $\gamma = (\gamma_1, \gamma_2, \gamma_3) \in \hat{\varsigma}^3$, the first sum been restricted to those γ for which $\gamma_1 = 1$ the second to those γ for which $\gamma_1 = -1$.

Choose a weight $\rho(k) \geq 0$, $k \in \tilde{Z}^d$ and consider the $\mathcal{G}$ valued process

$$p_t = \sum_{k \in \tilde{Z}^d} \rho(k)(x_k(t) \times \cos k.\theta + y_k(t) \times \sin k.\theta) \quad (3.2_b)$$

Consider the Stratonovitch SDE,

$$d\psi_t = -\left(\sum_{k \in \tilde{Z}^d} \rho(k) \left(\boldsymbol{\Gamma}(dx_k(t)) + \boldsymbol{\Gamma}(dy_k(t))\right)\right) \circ \psi_t = -\boldsymbol{\Gamma}(dp_t) \circ \psi_t, \quad (3.3)$$

$$\psi_0 = \text{Identity}.$$

As the $\boldsymbol{\Gamma}$ are antisymmetric operators this equation takes formally its values in the unitary group of $\mathcal{G}$; establishing this fact under mild assumptions on the weight ρ, will be the purpose of next four paragraphs.

The geometric meaning of (3.3) is to describe in terms of the algebraic parallelism inherited from the group structure of G the Levi-Civita parallelism inherited from the Riemannian structure of G; for this reason we call (3.3) the *equation of the stochastic parallel transport*.

Symmetries of the noise

Denote as before by G the group of measure preserving diffeomeophisms of the torus T^d and by $\mathcal{G}$ its Lie algebra.

Proposition 3.1. *Given $g \in G$, $z \in \mathcal{G}$, the adjoint action*

$$z \mapsto \frac{d}{d\epsilon}\bigg|_{\epsilon=0} g \exp(\epsilon z) g^{-1},$$

is the direct image $g_*(z)$ of the vector field z by the diffeomorphism g.

Proof. Reference [6], page 210.

In particular the translation $\tau_\varphi : \theta \mapsto \theta + \varphi$ is a diffeomorphism whose Jacobian matrix is the Identity; therefore

$$[(\tau_\varphi)_*(z)](\theta) = z(\theta - \varphi)$$

The collection $(\tau_\varphi)_*$, $\varphi \in T^d$, constitutes a unitary representation of T^d on $\mathcal{G}$ which decomposes into irreducible components along the basis

$$\cos k.\theta \otimes \mathcal{E}_k, \quad \sin k.\theta \otimes \mathcal{E}_k, \tag{3.4$_a$}$$

the action of $(\tau_\varphi)_*$ on $\mathcal{E}_k \oplus \mathcal{E}_k$ being the rotation

$$\mathcal{D}_k(\varphi) := \begin{pmatrix} \cos k\varphi & -\sin k\varphi \\ \sin k\varphi & \cos k\varphi \end{pmatrix}. \tag{3.4$_b$}$$

Furthermore τ_φ preserves the Lie algebra structure. The Christofell symbols are derived from the Hilbertian structure and from the bracket structure of $\mathcal{G}$. Therefore they commute with τ_φ in the sense that

$$(\tau_\varphi)_*[\Gamma(\xi)(\eta)] = \Gamma((\tau_\varphi)_*\xi)[(\tau_\varphi)_*\eta], \quad \xi, \eta \in \mathcal{G} \tag{3.5$_a$}$$

or denoting by $\boldsymbol{\Gamma}(z)$ the antihermitian endomorphism of $\mathcal{G}$ defined by the Christofell symbols, we have

$$\boldsymbol{\Gamma}((\tau_\varphi)_*(z)) = (\tau_\varphi)_* \circ \boldsymbol{\Gamma}(z) \circ (\tau_{-\varphi})_* \tag{3.5$_b$}$$

Denote by $\mathrm{su}(\mathcal{G})$ the vector space of antisymmetric operators on the Hilbert space $\mathcal{G}$. □

Proposition 3.2. *Let p_t the $\mathcal{G}$-valued process defined in (3.2$_b$) denote $p^\varphi =: (\tau_\varphi)_* p$; then p^φ and p have the same law.*

Proof. The rotation $\mathcal{D}_k(\phi)$ preserves in law the Brownian motion on $\mathcal{E}_k \otimes \mathcal{E}_k$. □

Corollary 3.1. *The processes $(\tau_\varphi) \circ \psi_t \circ (\tau_{-\varphi})$ and ψ_t have the same law.*

Proof. Denoting by ψ_t^p the solution of (3.3) associated the noise p_t, we have

$$(\tau_\varphi) \circ \psi_t \circ (\tau_{-\varphi}) = \psi_t^{p^\varphi}$$
□

Definition 3.1. We say that an endomorphism u of $\mathcal{G}$ is *pseudo diagonal* in the direct sum decomposition $\mathcal{G} = \bigoplus \mathcal{V}_k$ if the restriction of u to $\mathcal{V}_k$ takes its values in $\mathcal{V}_k$; then u is determined by a sequence $u_k \in End(\mathcal{V}_k)$; we denote $u = [u_k]$, $k \in \tilde{Z}^d$.

Theorem 3.1. *The Itô contraction describing the passage from the Stratonovitch equation* (3.3) *to its associated Itô form is an operator $\mathcal{B}$ which is pseudo diagonal in the direct sum decomposition*

$$\mathcal{G} = \bigoplus_{k \in \tilde{Z}^d} [\mathcal{E}_k \oplus \mathcal{E}_k] \tag{3.6$_a$}$$

Furthermore the diagonal term on the component $\mathcal{E}_k \oplus \mathcal{E}_k$ is of the form $-[\lambda_k, \lambda_k]$ where λ_k is a positive symmetric operator on $\mathcal{E}_k$.

Proof. Denote $\mathcal{E}_k^1$ (resp. $\mathcal{E}_k^2$) the first (resp. the second) component of the direct sum $\mathcal{E}_k \oplus \mathcal{E}_k$ and pick an orthonormal basis ϵ_k^α of each $\mathcal{E}_k$. As we have two injections of $\mathcal{E}_k \mapsto \mathcal{E}_k \oplus \mathcal{E}_k$, the first having its image in the first factor of the direct sum, the second in the second factor; we denote again ϵ_k^α the image by the first injection and η_k^α the image by the second injection; we get in this way an orthonormal basis of $\mathcal{E}_k^i$, $i = 1, 2$.

The Itô SDE has the shape

$$d\psi_t = -\left(\sum_{k \in \tilde{Z}^d} \rho \left(\mathbf{\Gamma}(dx_k(t)) + \mathbf{\Gamma}(dy_k(t)) \right) + \frac{\rho}{2} \mathbf{\Gamma}(dx_k) * \mathbf{\Gamma}(dx_k) + \frac{\rho}{2} \mathbf{\Gamma}(dy_k) * \mathbf{\Gamma}(dy_k) \right) \psi_t .$$

Denote by $x_k^\varphi(t) = (\tau_\varphi)_* x_k(t), y_k^\varphi(t) = (\tau_\varphi)_* y_k(t)$; then $x_k^\varphi(t)$ has the same law as $x_k(t)$. Therefore

$$(dx_t^\varphi) * (dx_t^\varphi) = (dx_t) * (dx_t)$$

Using formula (3.5$_b$),

$$\mathbf{\Gamma}(dx_t^\varphi) * \mathbf{\Gamma}(dx_t^\varphi) = (\tau_\varphi)_* \circ \mathbf{\Gamma}(dx) * \mathbf{\Gamma}(dx) \circ (\tau_{-\varphi})_*$$

Denoting

$$\mathcal{B} \, dt = \frac{1}{2} \sum_{k \in \tilde{Z}^d} [\rho(k)]^2 \left(\mathbf{\Gamma}(dx_k) * \mathbf{\Gamma}(dx_k) + \mathbf{\Gamma}(dy_k) * \mathbf{\Gamma}(dy_k) \right) \tag{3.6$_b$}$$

we have

$$\mathcal{B} = (\tau_\varphi)_* \circ \mathcal{B} \circ (\tau_{-\varphi})_* \tag{3.6$_c$}$$

Lemma 3.1. *An endomorphism of $\mathcal{G}$ satisfying* (3.6$_b$) *is diagonal in the basis* (3.6$_a$). □

Proof. Pick $A_s \in \mathcal{E}_s^1$, $A_k \in \mathcal{E}_k^1$, $B_s \in \mathcal{E}_s^2$, $B_k \in \mathcal{E}_k^2$ and consider the 2×2 matrix $M_{s,k}$

$$M_{s,k} := \begin{pmatrix} (A_s \mid \mathcal{B}(A_k)) & (B_s \mid \mathcal{B}(A_k)) \\ (A_s \mid \mathcal{B}(B_k)) & (B_s \mid \mathcal{B}(B_k)) \end{pmatrix}$$

Then formula (3.6_c) implies that

$$M_{s,k} = \mathcal{D}_s(\varphi) \circ M_{s,k} \circ \mathcal{D}_k(-\varphi) \qquad (3.6_d)$$

Integrating this identity over T^d we get

$$(2\pi)^d \, M_{s,k} = \int_{T^d} \mathcal{D}_s(\varphi) \circ M_{s,k} \circ \mathcal{D}_k(-\varphi) \, d\varphi_1 \otimes d\varphi_2$$

If we develop the product of matrices of the r.h.s. we find an expression of the form $(\cos s.\varphi)(\cos k.\varphi)$ which will have an integral equal to 0 if $s \neq k$.

As $\Gamma(*)$ are antisymmetric we get that their square are negative symmetric operators.

As on $\mathcal{E}_k \oplus \mathcal{E}_k$, this symmetric operator commutes with the rotations, it decomposes into two identical copies on each component. □

4. Transfer energy matrix of the stochastic parallel transport

A sequence of vectors with norm 1 of an Hilbert space can converge weakly to zero: this phenomena corresponds to a *"dissipation of energy towards the higher modes"*. For the construction of the stochastic parallel transport it is essential to control this dissipation.

We fix $\xi_0 \in \mathcal{G}$ and consider

$$\xi_t^p := \psi_t^p(\xi_0)$$

Pick $A_s \in \mathcal{E}_s^1$, $A_k \in \mathcal{E}_k^1$, $B_s \in \mathcal{E}_s^2$, $B_k \in \mathcal{E}_k^2$ and consider

$$\alpha_{s,k}^1(\xi_0) = E((A_s \mid \xi_t^p) \times (A_k \mid \xi_t^p)), \quad \alpha_{s,k}^2(\xi_0) = E((B_s \mid \xi_t^p) \times (B_k \mid \xi_t^p)),$$

$$\alpha_{s,k}^3(\xi_0) = E((A_s \mid \xi_t^p) \times (B_k \mid \xi_t^p))$$

Theorem 4.1. *Denote* $(\tau_\varphi)_*(\xi_0) := \xi_0^\varphi$, *then*

$$\int_{T^d} \alpha_{s,k}^i(\xi_0^\varphi) \, d\varphi = 0, \quad s \neq k, \ i = 1,2,3. \qquad (4.1)$$

Proof. Using (3.5$_b$) we get

$$\Gamma(p_t^\varphi) = (\tau_\varphi)_* \circ \Gamma(p_t) \circ (\tau_{-\varphi})_*$$

or by exponentiating

$$\psi_t^{p^\varphi} = (\tau_\varphi)_* \circ \psi_t^p \circ (\tau_{-\varphi})_*$$

$$(\tau_{-\varphi})_* \circ \psi_t^{p^\varphi} = \psi_t^p \circ (\tau_{-\varphi})_* \qquad (4.2_a)$$

Applying this identity to the vector ξ_0 and changing φ in $-\varphi$ we get

$$(\tau_\varphi)_* \circ \psi^{p_t^{-\varphi}} = \psi_t^p(\xi_0^\varphi)$$

Therefore

$$\alpha_{s,k}^1(\xi_0^\varphi) = E((A_s^{-\varphi} \mid \xi_t^{p^{-\varphi}}) \times (A_k^{-\varphi} \mid \xi_t^{p^{-\varphi}}))$$

Using now the key fact that $p_t^{-\varphi}$ and p_t have the same law, we obtain

$$\alpha_{s,k}^1(\xi_0^\varphi) = E((A_s^{-\varphi} \mid \xi_t^p) \times (A_k^{-\varphi} \mid \xi_t^p)) \qquad (4.2_b)$$

Using the identity $A_k^{-\varphi} = \cos(k.\varphi) \times A_k + \sin(k.\varphi) \times B_k$ and similar identities we get

$$\alpha_{s,k}^1(\xi_0^\varphi) = \cos(k.\varphi)\cos(s.\varphi)\alpha_{s,k}^1(\xi^0) + \sin(k.\varphi)\sin(s.\varphi)\alpha_{s,k}^2(\xi^0) \qquad (4.2_c)$$

$$+ \cos(k.\varphi)\sin(s.\varphi)\alpha_{s,k}^3(\xi^0) + \sin(k.\varphi)\cos(s.\varphi)\alpha_{k,s}^3(\xi^0) \qquad \square$$

Theorem 4.2. *There exists a matrix $\mathcal{A}$ defined on $\tilde{Z}^d \times \tilde{Z}^d$ such that denoting*

$$u_t^{\xi_0}(k) = E\left(\sum_{\alpha=1}^{d-1}(\epsilon_k^\alpha \mid \xi_t^p)^2 + (\eta_k^\alpha \mid \xi_t^p)^2\right),$$

where the basis ϵ_k^α, η_k^α, has been defined in (1.2), we have

$$\frac{du_t^{\xi_0}}{dt} = \mathcal{A}\, u_t^{\xi_0} \qquad (4.3_a)$$

Proof. We remark that (4.2$_a$) implies that

$$u_t^{\xi_0^\varphi}(k) = u_t^{\xi_0}(k) \quad \text{therefore}$$

$$\frac{1}{(2\pi)^d}\int_{T^d} u_t^{\xi_0^\varphi}(k)\, d\varphi = u_t^{\xi_0}(k) \qquad (4.3_b)$$

We can implement this fact by adding to the probability space generated by p a random initial value ξ_0^φ with a uniform repartition of the variable φ. By Itô calculus we can write

$$\frac{d}{dt}\int_{T^d} u_t^{\xi_0^\varphi}(k)$$

in terms of the

$$\int_{T^d} \alpha_{s,k}^i(\xi_0^\varphi)\, d\varphi_1 \otimes d\varphi_2$$

expression which vanishes for $s \neq k$ by (4.1). □

Corollary 4.1. *We have*

$$u_t^{\xi_0}(k) = 2E\bigg(\sum_{\alpha=1}^{d-1}(\epsilon_k^\alpha \mid \xi_t^p)^2\bigg), \qquad (4.3_c)$$

Proof. Use the identity (4.2_c) with $s = k$ and integrate in φ. □

5. Exact computation of the energy transfer matrix for $d = 2$

We shall present in section 6 indirect qualitative estimation of the energy transfer matrix valid in any dimension $d \geq 2$. In this section we recall the full computation from scratch of the transfer energy matrix in the case $d = 2$ which was obtained in Ref. 4. These computations give also another proof of the orthogonality results proved by symmetry arguments in section 4.

When $d = 2$ the space $\mathcal{E}_k$ is of dimension 1; we take for orthonormal basis of $\mathcal{E}_k$ the vector $(\frac{k_2}{|k|}, -\frac{k_1}{|k|})$. We get for orthonormal basis (for the L^2 metric) of vector fields with vanishing divergence on the torus, firstly the two constant vector fields and

$$A_k = \frac{1}{|k|}[(k_2 \cos k.\theta)\partial_1 - (k_1 \cos k.\theta)\partial_2)]$$

$$B_k = \frac{1}{|k|}[(k_2 \sin k.\theta)\partial_1 - (k_1 \sin k.\theta)\partial_2)]$$

where $k = (k_1, k_2) \in \tilde{Z}^2$, with $|k| \neq 0$, and where $k.\theta = k_1\theta_1 + k_2\theta_2$.

For $k, l \in \tilde{Z}^2$, $k, l \neq (0,0)$, define on $\tilde{Z}^2$ the following functions

$$\alpha_{k,l} := \frac{1}{4|k||l||k+l|}(|l+k|^2 - |k|^2 + |l|^2)) = \frac{1}{2|k||l||k+l|}((l \mid l+k))$$
(5.1$_a$)

$$\beta_{k,l} := \alpha_{-k,l} = \frac{1}{4|k||l||k-l|}(|l-k|^2 - |k|^2 + |l|^2)) = \frac{1}{2|k||l||k-l|}(l \mid (l-k))$$
(5.1$_b$)

$$[k, l] = k_1 l_2 - k_2 l_1.$$

The brackets of the above mentioned vector fields are given in the following

Theorem 5.1 (Ref. 4). *The following expressions hold*

$$[A_k, A_l] = \frac{[k,l]}{2|k||l|}(|k+l|B_{k+l} + |k-l|B_{k-l})$$

$$[B_k, B_l] = -\frac{[k,l]}{2|k||l|}(|k+l|B_{k+l} - |k-l|B_{k-l})$$

$$[A_k, B_l] = -\frac{[k,l]}{2|k||l|}(|k+l|A_{k+l} - |k-l|A_{k-l})$$

Concerning the Christoffel symbols,

Theorem 5.2 (Ref. 4). *The Cristoffel symbols are*

$$\Gamma_{A_k, A_l} = [k,l](\alpha_{k,l}B_{k+l} + \beta_{k,l}B_{k-l})$$

$$\Gamma_{B_k, B_l} = [k,l](-\alpha_{k,l}B_{k+l} + \beta_{k,l}B_{k-l})$$

$$\Gamma_{A_k, B_l} = [k,l](-\alpha_{k,l}A_{k+l} + \beta_{k,l}A_{k-l})$$

$$\Gamma_{B_k, A_l} = [k,l](-\alpha_{k,l}A_{k+l} - \beta_{k,l}A_{k-l})$$

In this two-dimensional case the equation for the stochastic parallel transport reads

$$d\psi_t = -\left(\sum_{k \neq 0} \rho(k)\Gamma_{A_k,.}\circ dx^k + \rho(k)\Gamma_{B_k,.}\circ dy^k\right)\psi_t$$

with independent Brownian motions in all components, $\psi(0) = id$.

We remark the difference with respect to Ref. 4 of the constants appearing in (5.1$_a$) and (5.1$_b$): taking indices in $\tilde{Z}^2$, only half of the quantities computed in Ref. 4, Theorem 3.1, have to be considered.

Theorem 5.3 (Ref. 4). *The coefficient of the transfer energy matrix are given by*

$$\mathcal{A}_l^l = -2\sum_k [\rho(k)]^2 [l,k]^2 \times \left(\beta_{k,k-l}\beta_{k,l} + \alpha_{k,-k-l}\alpha_{k,l} \right)$$

$$\mathcal{A}_s^l = 2\sum_k [\rho(k)]^2 [l,k]^2 \times \left(\alpha_{k,l-k}^2 \, \delta_s^{l-k} + \beta_{k,l+k}^2 \, \delta_s^{l+k} \right)$$

The matrix $\mathcal{A}$ is symmetric, the non diagonal terms are positive, the diagonal terms are negative, the sum of the coefficients on each column vanishes.

Remark 5.1. Notice that

$$\beta_{k,k-l}\beta_{k,l} = \alpha_{k,l-k}^2 = \frac{(l|l-k)^2}{4|k|^2|l|^2|l-k|^2}$$

$$\alpha_{k,-k-l}\alpha_{k,l} = \beta_{k,l+k}^2 = \frac{(l|l+k)^2}{4|k|^2|l|^2|l+k|^2}$$

Corollary 5.1. *Denote $\chi_{k,l}$ the angle between the vectors k and l; then*

$$|\mathcal{A}_l^l| = \frac{|l|^2}{8} \sum_{k \in \tilde{Z}^2} (1 - \cos 4\chi_{k,l}) \times [\rho(k)]^2$$

$$\sum_k [\rho(k)]^2 [l,k]^2 \times k \left(\alpha_{k,l-k}^2 - \beta_{k,l+k}^2 \right) = 0$$

If $\rho(k)$ depends only upon $|k|$

$$|\mathcal{A}_l^l| \simeq c|l|^2, \quad 4c := \sum_{k \in \tilde{Z}^2} [\rho(k)]^2 \qquad (5.2)$$

6. Qualitative estimation of the energy transfer matrix for d > 2

We propose ourselves to obtain in dimension greater than 2 the estimates (5.2).

We interpreted the ${}^\gamma \Gamma_{*,s}^l$ as defining operators $\mathcal{E}_\cdot \to \mathcal{E}_l$, the euclidean structure of the $\mathcal{E}_j$ makes possible to define their adjoint; then

$$\text{trace}({}^\gamma \Gamma_{*,s}^l)({}^\gamma \Gamma_{*,s}^l)^*$$

is well defined and is equal to the HS norm (Hilbert-Schmidt norm).

Theorem 6.1. *The diagonal terms of the matrix $\mathcal{A}$ are expressed by the diagonal terms of the diagonal matrix $\mathcal{B}$ defined in* (3.6$_a$),

$$\mathcal{A}_l^l = -2 \; trace \; (\lambda_k) \tag{6.1$_a$}$$

and, as a consequence, are negative; the non diagonal terms are given by

$$\mathcal{A}_s^l = \sum_{k \in \tilde{Z}^d} [\rho(k)]^2 \left| \sum_{\gamma_1=1} \gamma \Gamma_{k,s}^l \right|_{HS}^2 + \sum_{k \in \tilde{Z}^d} [\rho(k)]^2 \left| \sum_{\gamma_1=-1} \gamma \Gamma_{k,s}^l \right|_{HS}^2, \quad l \neq s; \tag{6.1$_b$}$$

The non diagonal terms are positive. The sum of the terms of each column vanishes. The matrix $\mathcal{A}$ is symmetric.

Theorem 6.2.

$$\frac{1}{4(d-1)^2} \mathcal{A}_l^l \leq |l|^2 \sum_k [\rho(k)]^2 + \sum_k |k|^2 [\rho(k)]^2. \tag{6.2}$$

Proof. We have,

$$|\mathcal{A}_l^l| \leq \sum_{k,s} \|\Gamma_{k,s}^l\|_{HS}^2 [\rho(k)]^2$$

which by (2.4$_a$) is equal to

$$\sum_k [\rho(k)]^2 (\|b_{k,l-k}\|_{HS}^2 + \|b_{k-l,l}\|_{HS}^2 + \|b_{l+k,k}\|_{HS}^2)$$

Finally, using the expression of $b_{k,j}$ we get the inequality $\|b_{k,j}\|_{HS}^2 \leq (d-1)^2(|k|^2 + |j|^2)$.

□

References

1. H. Airault, P. Malliavin. Quasi-invariance of Brownian measures on the group of circle homeomorphisms and infinite-dimensional Riemannian geometry. *J. Funct. Anal.* **241** (1) (2006), 99–142.
2. V. I. Arnold. Sur la géométrie différentielle des groupes de Lie de dimension infinie et ses applications a l'hidrodynamique des fluides parfaits. *Ann. Inst. Fourier* **16** (1966), 316–361.
3. V. I. Arnold, B. A. Khesin. *Topological Methods in Hydrodynamics.* Springer-Verlag, New York, 1998.
4. A. B. Cruzeiro, F. Flandoli, P. Malliavin. Brownian motion on volume preserving diffeomorphisms group and existence of global solutions of 2D stochastic Euler equation. *J. Funct. Anal.* **242** (2007), 304–326.
5. D. Ebin, J. Marsden. Groups of diffeomorphisms and the motion of incompressible fluid. *Ann. Math.* **92** (2) (1970), 102–163.

6. P. Malliavin. *Stochastic Analysis*. Grund. der Mathem. Wissen., Springer-Verlag, 1997.

19

Riemannian geometry of $\mathrm{Diff}(S^1)/S^1$ revisited

Maria Gordina*
*Department of Mathematics
University of Connecticut
Storrs, CT 06269, USA
E-mail: gordina@math.uconn.edu*

A further study of Riemannian geometry $\mathrm{Diff}(S^1)/S^1$ is presented. We describe Hermitian and Riemannian metrics on the complexification of the homogeneous space, as well as the complexified symplectic form. It is based on the ideas from Ref. 12, where instead of using the Kähler structure symmetries to compute the Ricci curvature, the authors rely on classical finite-dimensional results of Nomizu et al on Riemannian geometry of homogeneous spaces.

Keywords: Virasoro algebra, group of diffeomorphisms, Ricci curvature

1. Introduction

Let $\mathrm{Diff}(S^1)$ be the Virasoro group of orientation-preserving diffeomorphisms of the unit circle. Then the quotient space $\mathrm{Diff}(S^1)/S^1$ describes those diffeomorphisms that fix a point on the circle. The geometry of this infinite-dimensional space has been of interest to physicists (e.g. Refs. 7,8,19).

We follow the approach taken in Ref. 7,8,14,19 in that we describe the space $\mathrm{Diff}(S^1)/S^1$ as an infinite dimensional complex manifold with a Kähler metric. Theorem 3.1 describes properties of the Hermitian and Riemannian metrics, as well as of the complexified symplectic form. Then we introduce the covariant derivative ∇ which is consistent with the Kähler structure. We use the expression for the derivative found in Ref. 12, where the classical finite-dimensional results of K. Nomizu[16] for homogeneous spaces were used in this infinite-dimensional setting. The goal of the present article is to clarify certain parts of Ref. 12, in particular, Theorem 4.5. This theorem stated that the covariant derivative in question is Levi-Civita, but

---

*The research of the author is partially supported by the NSF Grant DMS-0306468 and the Humboldt Foundation Research Fellowship.

the details were omitted. In the present paper we explicitly define the Riemannian metric g for which ∇ is the Levi-Civita covariant derivative. This is proven in part (3) of Theorem 3.2 of the present paper. To complete the exposition we present the computation of the Riemannian curvature tensor and the Ricci curvature for the covariant derivative ∇. This proof follows the one in Ref. 12.

Our interest to the geometry of this infinite-dimensional manifold comes from attempts to develop stochastic analysis on infinite-dimensional manifolds. Relevant references include works by H. Airault, V. Bogachev, P. Malliavin, A. Thalmaier.[2-6,10] A group Brownian motion in Diff(S^1) has been constructed by P. Malliavin.[15] From the finite-dimensional case we know that the lower bound of the Ricci curvature controls the growth of the Brownian motion, therefore a better understanding of the geometry of Diff(S^1)/S^1 might help in studying a Brownian motion on this homogeneous space. For further references to the works exploring the connections between stochastic analysis and Riemannian geometry in infinite dimensions, mostly in loop groups and their extensions such as current groups, path spaces and complex Wiener spaces see Refs. 9,11,17,18.

2. Virasoro algebra

Notation 2.1. Let Diff(S^1) be the group of orientation preserving C^∞-diffeomorphisms of the unit circle, and diff(S^1) its Lie algebra. The elements of diff(S^1) will be identified with the C^∞ left-invariant vector fields $f(t)\frac{d}{dt}$, with the Lie bracket given by

$$[f,g] = fg' - f'g, f, g \in \text{diff}(S^1).$$

Definition 2.1. Suppose c, h are positive constants. Then the **Virasoro algebra** $\mathcal{V}_{c,h}$ is the vector space $\mathbb{R} \oplus \text{diff}(S^1)$ with the Lie bracket given by

$$[a\kappa+f, b\kappa+g]_{\mathcal{V}_{c,h}} = \omega_{c,h}(f,g)\kappa + [f,g], \quad (1)$$

where $\kappa \in \mathbb{R}$ is the central element, and ω is the bilinear symmetric form

$$\omega_{c,h}(f,g) = \int_0^{2\pi} \left((2h - \frac{c}{12}) f'(t) - \frac{c}{12} f^{(3)}(t) \right) g(t) \frac{dt}{2\pi}.$$

Remark 2.1. If $h = 0$, $c = 6$, then $\omega_{c,h}$ is the fundamental cocycle ω (see Ref. 3)

$$\omega(f,g) = -\int_0^{2\pi} \left(f' + f^{(3)} \right) g \frac{dt}{4\pi}.$$

Remark 2.2. A simple verification shows that $\mathcal{V}_{c,h}$ with $\omega_{c,h}$ satisfies the Jacobi identity, and therefore $\mathcal{V}_{c,h}$ with this bracket is indeed a Lie algebra. In addition, by the integration by parts formula $\omega_{c,h}$ satisfies

$$\omega_{c,h}\left(f', g\right) = -\omega_{c,h}\left(f, g'\right). \tag{2}$$

Moreover, $\omega_{c,h}$ is anti-symmetric

$$\omega_{c,h}\left(f, g\right) = -\omega_{c,h}\left(g, f\right). \tag{3}$$

Notation 2.2. Throughout this work we use $k, m, n... \in \mathbb{N}$, and $\alpha, \beta, \gamma... \in \mathbb{Z}$.

Below we introduce an inner product on the Lie algebra $\text{diff}(S^1)$ which has a natural basis

$$f_k = \cos kt, g_m = \sin mt, \ k = 0, 1, 2..., m = 1, 2.... \tag{4}$$

The Lie bracket in this basis satisfies the following identities

$$[f_m, f_n] = \frac{1}{2}\left((m-n)g_{m+n} + (m+n)\frac{m-n}{|m-n|}g_{|m-n|}\right), \ m \neq n,$$

$$[g_m, g_n] = \frac{1}{2}\left((n-m)g_{m+n} + (m+n)\frac{m-n}{|m-n|}g_{|m-n|}\right), \ m \neq n, \tag{5}$$

$$[f_m, g_n] = \frac{1}{2}\left((n-m)f_{m+n} + (m+n)f_{|m-n|}\right).$$

Notation 2.3. By $\text{diff}_0(S^1)$ we denote the space of functions having mean 0. This space can be identified with $\text{diff}(S^1)/S^1$, where S^1 is being viewed as constant vector fields corresponding to rotations of S^1.

Then any element of $f \in \text{diff}_0(S^1)$ can be written

$$f(t) = \sum_{k=1}^{\infty}\left(a_k f_k + b_k g_k\right),$$

with $\{a_k\}_{k=1}^{\infty}, \{b_k\}_{k=1}^{\infty} \in \ell^2$ since f is smooth. We will also need the following endomorphism J of $\text{diff}_0(S^1)$

$$J(f)(t) = \sum_{k=1}^{\infty}\left(b_k f_k - a_k g_k\right). \tag{6}$$

It satisfies $J^2 = -I$.

Notation 2.4. For any $k \in \mathbb{Z}$ we denote $\theta_k = 2hk + \frac{c}{12}(k^3 - k)$.

Remark 2.3. Note that $\theta_{-k} = -\theta_k$, for any $k \in \mathbb{Z}$.

The form $\omega_{c,h}$ and the endomorphism J induce an inner product on $\mathrm{diff}_0(S^1)$ by

$$\langle f, g \rangle = \omega_{c,h}(f, Jg) = \omega_{c,h}(g, Jf).$$

The last identity follows from Equation (3).

Proposition 2.1. $\langle f, g \rangle$ *is an inner product on* $\mathrm{diff}_0(S^1)$.

Proof. Let $b_0 = 0$, then

$$\omega_{c,h}(f, Jf) = \int_0^{2\pi} \left(\left(2h - \frac{c}{12}\right) f'(t) - \frac{c}{12} f^{(3)}(t) \right) (Jf)(t) \frac{dt}{2\pi} =$$

$$\int_0^{2\pi} \left(\sum_{k=1}^{\infty} \theta_k (b_k f_k - a_k g_k) \right) \left(\sum_{m=1}^{\infty} (b_m f_m - a_m g_m) \right) \frac{dt}{2\pi} = \frac{1}{2} \sum_{k=1}^{\infty} \theta_k (a_k^2 + b_k^2).$$

Then for any $f \in \mathrm{diff}_0(S^1)$

$$\langle f, f \rangle = \frac{1}{2} \sum_{k=1}^{\infty} \theta_k \left(a_k(f)^2 + b_k(f)^2 \right). \qquad \square$$

Notation 2.5. Let $\lambda_{m,n} = \frac{(2n+m)\theta_m}{2\theta_{m+n}}$ for any $n, m \in \mathbb{Z}$. Then it is easy to check that

$$\lambda_{m,n} = \lambda_{n,m} + \frac{m-n}{2}. \tag{7}$$

3. Diff$(S^1)/S^1$ as a Kähler manifold

Denote $\mathfrak{g} = \mathrm{diff}(S^1)$, $\mathfrak{m} = \mathrm{diff}_0(S^1)$, $\mathfrak{h} = f_0 \mathbb{R}$, so that $\mathfrak{g} = \mathfrak{m} \oplus \mathfrak{h}$. Then $\mathfrak{g}$ is an infinite-dimensional Lie algebra equipped with an inner product $\langle \cdot, \cdot \rangle$. Note that for any $n \in \mathbb{N}$

$$[f_0, f_n] = -ng_n \in \mathfrak{m}, \quad [g_0, g_n] = nf_n \in \mathfrak{m},$$

and therefore $[\mathfrak{h}, \mathfrak{m}] \subset \mathfrak{m}$. In addition, $\mathfrak{h}$ is a Lie subalgebra of $\mathfrak{g}$, but $\mathfrak{m}$ is not a Lie subalgebra of $\mathfrak{g}$ since $[f_m, g_m] = m f_0$.

Let $G = \mathrm{Diff}(S^1)$ with the associated Lie algebra $\mathrm{diff}(S^1)$, the subgroup $H = S^1$ with the Lie algebra $\mathfrak{h} \subset \mathfrak{g}$, then $\mathfrak{m}$ is a tangent space naturally associated with the quotient $\mathrm{Diff}(S^1)/S^1$. For any $g \in \mathfrak{g}$ we denote by $g_{\mathfrak{m}}$ (respectively $g_{\mathfrak{h}}$) its $\mathfrak{m}$-(respectively $\mathfrak{h}$-)component, that is, $g = g_{\mathfrak{m}} + g_{\mathfrak{h}}$, $g_{\mathfrak{m}} \in \mathfrak{m}$, $g_{\mathfrak{h}} \in \mathfrak{h}$. The fact $[\mathfrak{h}, \mathfrak{m}] \subset \mathfrak{m}$ implies that for any $h \in \mathfrak{h}$ the adjoint representation $ad(h) = [h, \cdot] : \mathfrak{g} \to \mathfrak{g}$ maps $\mathfrak{m}$ into $\mathfrak{m}$. We will abuse notation by using $ad(h)$ for the corresponding endomorphism of $\mathfrak{m}$.

Recall that $J : \text{diff}_0(S^1) \to \text{diff}_0(S^1)$ is an endomorphism defined by (6), or equivalently, in the basis $\{f_m, g_n\}$, $m, n = 1, \ldots$ by

$$Jf_m = -g_m, \quad Jg_n = f_n.$$

This is an almost complex structure on $\text{diff}_0(S^1)$, and as was shown in Ref. 12 it is actually a complex structure for an appropriately chosen connection.

Let $\mathfrak{g}_\mathbb{C}$ and $\mathfrak{m}_\mathbb{C}$ be the complexifications of $\mathfrak{g}$ and $\mathfrak{m}$ respectively. Now we would like to introduce Hermitian metric, Riemannian metric and the complexified symplectic form $\omega_\mathbb{C}$.

Notation 3.1. For any $f + ig, u + iv \in \mathfrak{g}_\mathbb{C}$, where $f, g, u, v \in \mathfrak{g}$ and $i^2 = -1$ we denote

$$h(f + ig, u + iv) = \langle f, u \rangle + \langle g, v \rangle + i(\langle g, u \rangle - \langle f, v \rangle);$$
$$g(f + ig, u + iv) = \langle f, u \rangle + \langle g, v \rangle = \text{Re}(h(f + ig, u + iv));$$
$$\omega_\mathbb{C}(f + ig, u + iv) = \langle g, u \rangle - \langle f, v \rangle = \text{Im}(h(f + ig, u + iv)).$$

We will call h a Hermitian metric, g a Riemannian metric, and $\omega_\mathbb{C}$ a symplectic form.

The endomorphism J can be naturally extended to $\mathfrak{g}_\mathbb{C}$ by $J(f + ig) = Jf + iJg$ for any $f, g \in \mathfrak{g}$. We will abuse notation and use the same J for this extended endomorphism. It is easy to check that J is complex-linear.

Notation 3.2. Let $m \in \mathbb{N}$, then denote

$$L_m = f_m + ig_m, \quad L_{-m} = f_m - ig_m.$$

One can see that on elements $\{L_\alpha\}_{\alpha \in \mathbb{Z}}$ the endomorphism J acts in the following way

$$J(L_\alpha) = i\,\text{sgn}(\alpha)L_\alpha, \alpha \in \mathbb{Z}.$$

Using Equation(5) we can easily check that for any $\alpha, \beta \in \mathbb{Z}$

$$[L_\alpha, L_\beta] = i(\beta - \alpha)L_{\alpha + \beta}. \tag{8}$$

In addition to these properties, we can easily see that $\{L_\alpha\}_{\alpha \in \mathbb{Z}}$ form an orthogonal system in the Riemannian metric g. Indeed, for any $m, n \in \mathbb{N}$

$$h(L_m, L_n) = h(f_m + ig_m, f_n + ig_n) =$$
$$\langle f_m, f_n \rangle + \langle g_m, g_n \rangle = \theta_m \delta_{m,n} = g(L_m, L_n);$$
$$h(L_m, L_{-n}) = h(L_{-m}, L_n) = 0.$$

In particular,
$$\omega_{\mathbb{C}}(L_\alpha, L_\beta) = 0$$
for any $\alpha, \beta \in \mathbb{Z}$.

Theorem 3.1. *For any $F, G \in \mathfrak{g}_{\mathbb{C}}$, $a, b \in \mathbb{C}$ we have*

(1) $h(F,G) = \overline{h(G,F)}$, $g(F,G) = g(G,F)$, $\omega_{\mathbb{C}}(F,G) = -\omega_{\mathbb{C}}(G,F)$;
(2) $h(aF, bG) = a\bar{b}h(F, G)$,
$g(aF, bG) = \operatorname{Re}(a\bar{b}) g(F,G) - \operatorname{Im}(a\bar{b}) \omega_{\mathbb{C}}(F, G)$,
$\omega_{\mathbb{C}}(aF, bG) = \operatorname{Re}(a\bar{b}) \omega_{\mathbb{C}}(F, G) + \operatorname{Im}(a\bar{b}) g(F, G)$;
(3) Both the Hermitian and Riemannian inner products, as well as the form $\omega_{\mathbb{C}}$ are invariant under J;
(4) the form $\omega_{\mathbb{C}}$ is closed.

Proof. Parts (1) and (2) follow from the definition of the metrics h and g and form $\omega_{\mathbb{C}}$ by a straightforward computation. Let us check the invariance of h, g and $\omega_{\mathbb{C}}$ under J and closedness of $\omega_{\mathbb{C}}$. It is enough to check that $h(JF, JF) = h(F, F)$ on the complex basis $\{f_m, if_m, g_n, g_n\}$. This is indeed so for $\{f_m, g_n\}$, and for $F = if_m$ by the second part of this Proposition

$$h(J(if_m), J(if_m)) = h(f_m, f_m) = h(if_m, if_m).$$

The rest of the identities can be checked similarly, and invariance of the Hermitian metric under J obviously implies invariance of g and $\omega_{\mathbb{C}}$.

To check closedness of $\omega_{\mathbb{C}}$ we can first note that (2) holds on $\mathfrak{g}_{\mathbb{C}}$ as well, and so

$$d\omega_{\mathbb{C}}(F, G) = d\omega_{\mathbb{C}}(F', G) + d\omega_{\mathbb{C}}(F, G') = 0. \qquad \square$$

Notation 3.3. Define a complex-linear connection on $\mathfrak{m}_{\mathbb{C}}$ by

$$\begin{aligned}
\nabla_{L_m} L_n &= -2i\lambda_{m,n} L_{m+n}; \\
\nabla_{L_{-m}} L_{-n} &= 2i\lambda_{m,n} L_{-m-n}; \\
\nabla_{L_{-m}} L_n &= i(m+n) L_{n-m}, \ n > m; \\
\nabla_{L_m} L_{-n} &= -i(m+n) L_{m-n}, \ n > m; \\
\nabla_{L_{-m}} L_n &= \nabla_{L_m} L_{-n} = 0, \ m \geqslant n,
\end{aligned} \qquad (9)$$

for any $m, n \in \mathbb{N}$.

As we can see from (9), for any $\alpha, \beta \in \mathbb{Z}$

$$\nabla_{L_\alpha} L_\beta = \Gamma_{\alpha, \beta} L_{\alpha + \beta}$$

for some $\Gamma_{\alpha,\beta} \in \mathbb{C}$. In addition, it is easy to check that the definition of the covariant derivative ∇ in (9) implies that

$$\Gamma_{-\alpha,-\beta} = -\Gamma_{\alpha,\beta}, \qquad (10)$$

and (9) can be described as

$$\Gamma_{m,n} = -2i\lambda_{m,n}, \ \Gamma_{-m,n} = i(m+n), n > m, \ \Gamma_{-m,n} = 0, n \leqslant m,$$
$$\Gamma_{-\alpha,-\beta} = -\Gamma_{\alpha,\beta}, \ m, n \in \mathbb{N}, \ \alpha, \beta \in \mathbb{Z}.$$

The first part of the following theorem is an analogue of the Newlander-Nirenberg theorem in our setting.

Theorem 3.2. *The covariant derivative ∇ and the almost complex structure J satisfy the following properties.*

(1) The Nijenhuis tensor N (the torsion of the almost complex structure J) defined by

$$N_J(X,Y) = 2\left([JX, JY]_{\mathfrak{m}_{\mathbb{C}}} - [X,Y]_{\mathfrak{m}_{\mathbb{C}}} - J[X, JY]_{\mathfrak{m}_{\mathbb{C}}} - J[JX, Y]_{\mathfrak{m}_{\mathbb{C}}}\right)$$

vanishes on $\mathfrak{m}_{\mathbb{C}} = \mathrm{diff}_0(S^1)_{\mathbb{C}}$;

(2) J is a complex structure on $\mathfrak{m}_{\mathbb{C}} = \mathrm{diff}_0(S^1)_{\mathbb{C}}$ with the covariant derivative ∇ defined in Notation 9. In other words, J is integrable;

(3) the covariant derivative ∇ is the Levi-Civita connection, that is, it is torsion-free and compatible with the Riemannian metric g.

Proof.

(1) It is enough to check the statement on the real basis elements.

$$N_J(L_\alpha, L_\beta) =$$
$$2\left([JL_\alpha, JL_\beta]_{\mathfrak{m}_{\mathbb{C}}} - [L_\alpha, L_\beta]_{\mathfrak{m}_{\mathbb{C}}} - J\left([L_\alpha, JL_\beta]_{\mathfrak{m}_{\mathbb{C}}}\right) - J\left([JL_\alpha, L_\beta]_{\mathfrak{m}_{\mathbb{C}}}\right)\right) =$$
$$-2((1 + \mathrm{sgn}(\alpha)\mathrm{sgn}(\beta))[L_\alpha, L_\beta]_{\mathfrak{m}_{\mathbb{C}}} + i(\mathrm{sgn}(\alpha) + \mathrm{sgn}(\beta))J([L_\alpha, L_\beta]_{\mathfrak{m}_{\mathbb{C}}})) =$$
$$0.$$

Here we used Equation (8) for the Lie bracket $[L_\alpha, L_\beta]_{\mathfrak{m}_{\mathbb{C}}}$, and then considered the cases when α and β are of the same or opposite signs.

(2) We will use the fact that

$$(\nabla_X J)(Y) = \nabla_X(JY) - J(\nabla_X Y).$$

One can check using Equation (9) that $J(\nabla_{L_\alpha} L_\beta) = i\,\mathrm{sgn}(\beta)\nabla_{L_\alpha} L_\beta$. Then again on the real basis elements

$$(\nabla_{L_\alpha} J)(L_\beta) = \nabla_{L_\alpha}(JL_\beta) - J(\nabla_{L_\alpha} L_\beta) =$$
$$i\,\mathrm{sgn}(\beta)\nabla_{L_\alpha} L_\beta - J(\nabla_{L_\alpha} L_\beta) = 0$$

since the covariant derivative $\nabla_{L_\alpha} L_\beta$ is again a basis element.

(3) The torsion of the covariant derivative ∇ is defined by

$$T_\nabla(X,Y) = \nabla_X Y - \nabla_Y X - [X,Y]_{\mathfrak{m}_{\mathbb{C}}}.$$

Using Equation (7) and the definition of the covariant derivative in Notation 9 we see that

$$\nabla_{L_m} L_n - \nabla_{L_n} L_m = 2i\,(\lambda_{n,m} - \lambda_{m,n})\,L_{m+n} = i(n-m)L_{m+n};$$
$$\nabla_{L_{-m}} L_{-n} - \nabla_{L_{-n}} L_{-m} = 2i\,(\lambda_{m,n} - \lambda_{n,m})\,L_{-m-n} = i(m-n)L_{-m-n};$$
$$\nabla_{L_{-m}} L_n - \nabla_{L_n} L_{-m} = i(m+n)L_{n-m},\ n > m;$$
$$\nabla_{L_{-m}} L_n - \nabla_{L_n} L_{-m} = i(m+n)L_{n-m},\ n < m;$$
$$\nabla_{L_{-n}} L_n - \nabla_{L_n} L_{-n} = \nabla_{L_n} L_{-n} - \nabla_{L_{-n}} L_n = 0;$$
$$\nabla_{L_m} L_{-n} - \nabla_{L_{-n}} L_m = -i(m+n)L_{m-n},\ n > m;$$
$$\nabla_{L_m} L_{-n} - \nabla_{L_{-n}} L_m = -i(m+n)L_{m-n},\ n < m,$$

which together with the Lie bracket expression for the basis elements $\{L_\alpha\}$ in Equation(8) gives the result.

To check that ∇ is compatible with the Riemannian metric g, it is enough to check that

$$g(\nabla_{L_\alpha} L_\beta, L_\gamma) + g(L_\beta, \nabla_{L_\alpha} L_\gamma) = 0$$

for any $L_\alpha, L_\beta, L_\gamma \in \mathbb{Z}$. Note that all $\Gamma_{\delta,\varepsilon}$ are either 0 or purely imaginary, so

$$g(\nabla_{L_\alpha} L_\beta, L_\gamma) + g(L_\beta, \nabla_{L_\alpha} L_\gamma) =$$
$$g(\Gamma_{\alpha,\beta} L_{\alpha+\beta}, L_\gamma) + g(L_\beta, \Gamma_{\alpha,\gamma} L_{\alpha+\gamma}) =$$
$$\Gamma_{\alpha,\beta}\omega_{\mathbb{C}}(L_{\alpha+\beta}, L_\gamma) - \Gamma_{\alpha,\gamma}\omega_{\mathbb{C}}(L_\beta, L_{\alpha+\gamma}) = 0. \qquad \square$$

Definition 3.1. The **curvature tensor** $R_{xy}: \mathfrak{m}_{\mathbb{C}} \to \mathfrak{m}_{\mathbb{C}}$ is defined by

$$R_{xy} = \nabla_x \nabla_y - \nabla_y \nabla_x - \nabla_{[x,y]_{\mathfrak{m}_{\mathbb{C}}}} - ad([x,y]_{\mathfrak{h}_{\mathbb{C}}}),\ x,y \in \mathfrak{m}_{\mathbb{C}};$$

the **Ricci tensor** $\mathrm{Ric}(x,y): \mathfrak{m}_{\mathbb{C}} \to \mathfrak{m}_{\mathbb{C}}$ is defined by

$$\mathrm{Ric}(x,y) = \sum_{n \in \mathbb{N}} \frac{1}{\theta_n} \left(\langle R_{L_n,x} y, L_n\rangle + \langle R_{L_{-n},x} y, L_{-n}\rangle\right).$$

Proposition 3.1. *For any $\alpha, \beta, \gamma \in \mathbb{Z}$*

$$R_{L_\alpha, L_\beta} L_\gamma = R_{\alpha,\beta,\gamma} L_{\alpha+\beta+\gamma},$$

where the coefficients $R_{\alpha,\beta,\gamma}$ satisfy

$$R_{-\alpha,-\beta,-\gamma} = \overline{R_{\alpha,\beta,\gamma}}.$$

Proof.

$$R_{L_\alpha, L_\beta} L_\gamma = \nabla_{L_\alpha} \nabla_{L_\beta} L_\gamma - \nabla_{L_\beta} \nabla_{L_\alpha} L_\gamma - i(\beta - \alpha) \nabla_{L_{\alpha+\beta}} L_\gamma =$$
$$\left(\Gamma_{\beta,\gamma} \Gamma_{\alpha,\beta+\gamma} - \Gamma_{\alpha,\gamma} \Gamma_{\beta,\alpha+\gamma} - i(\beta - \alpha) \Gamma_{\alpha+\beta,\gamma} \right) L_{\alpha+\beta+\gamma}.$$

Then Equation(10) implies

$$R_{-\alpha,-\beta,-\gamma} = \Gamma_{-\beta,-\gamma} \Gamma_{-\alpha,-\beta-\gamma} - \Gamma_{-\alpha,-\gamma} \Gamma_{-\beta,-\alpha-\gamma} - i(\alpha - \beta) \Gamma_{-\alpha-\beta,-\gamma} = \overline{R_{\alpha,\beta,\gamma}}.$$ □

Theorem 3.3. *[Theorem 4.11 in Ref. 12] The only non-zero components of the Ricci tensor are*

$$\operatorname{Ric}\left(\frac{L_n}{\sqrt{|\theta_n|}}, \frac{L_{-n}}{\sqrt{|\theta_n|}} \right) = -\frac{13n^3 - n}{6\theta_n}, \quad n \in \mathbb{Z}.$$

Proof.

$$\operatorname{Ric}\left(\frac{L_\alpha}{\sqrt{|\theta_\alpha|}}, \frac{L_\beta}{\sqrt{|\theta_\beta|}} \right) = \sum_{m \in \mathbb{N}} \frac{\left(\langle R_{L_m, L_\alpha} L_\beta, L_m \rangle + \langle R_{L_{-m}, L_\alpha} L_\beta, L_{-m} \rangle \right)}{\sqrt{|\theta_\alpha|} \sqrt{|\theta_\beta|} \theta_m} =$$
$$\delta_{\alpha,-\beta} \sum_{m \in \mathbb{N}} \frac{\left(\langle R_{L_m, L_\alpha} L_{-\alpha}, L_{-m} \rangle + \langle R_{L_{-m}, L_\alpha} L_{-\alpha}, L_m \rangle \right)}{|\theta_\alpha| \theta_m}.$$

Thus the only non-zero components of the Ricci tensor are the ones when $\alpha + \beta = 0$. For example, let $\alpha = n \in \mathbb{N}$, then according to Equation (9) we have that

$$R_{L_m, L_n} L_{-n} = \nabla_{L_m} \nabla_{L_n} L_{-n} - \nabla_{L_n} \nabla_{L_m} L_{-n} - \nabla_{[L_m, L_n]_{\mathfrak{m}}} L_{-n} - \operatorname{ad}([L_m, L_n]_{\mathfrak{h}})$$
$$L_{-n} = -\nabla_{L_n} \nabla_{L_m} L_{-n} - i(n - m) \nabla_{L_{m+n}} L_{-n} = -\nabla_{L_n} \nabla_{L_m} L_{-n},$$

therefore

$$R_{L_m, L_n} L_{-n} = 0, \quad m \geqslant n;$$
$$R_{L_m, L_n} L_{-n} = i(m+n) \nabla_{L_n} L_{m-n} = 0, \quad m < n.$$

Now by Proposition 3.1

$$\begin{aligned}
R_{L_m, L_{-n}} L_n &= R_{L_{-m}, L_n} L_{-n} \\
&= \nabla_{L_{-m}} \nabla_{L_n} L_{-n} - \nabla_{L_n} \nabla_{L_{-m}} L_{-n} \\
&\quad - \nabla_{[L_{-m}, L_n]_{\mathfrak{m}}} L_{-n} - \mathrm{ad}([L_{-m}, L_n]_{\mathfrak{h}}) L_{-n} \\
&= -\nabla_{L_n} \nabla_{L_{-m}} L_{-n} - i(m+n) \nabla_{L_{n-m}} L_{-n} \\
&= -2i\lambda_{m,n} \nabla_{L_n} L_{-m-n} - i(m+n) \nabla_{L_{n-m}} L_{-n} \\
&= -2(m+2n) \lambda_{m,n} L_{-m} - i(m+n) \nabla_{L_{n-m}} L_{-n},
\end{aligned}$$

thus

$$R_{L_{-m}, L_n} L_{-n} = -2(m+2n) \lambda_{m,n} L_{-m} + 2(m+n) \lambda_{m-n,n} L_{-m}, \quad m > n;$$
$$R_{L_{-m}, L_n} L_{-n} = -\left(2(m+2n) \lambda_{m,n} + (2n-m)(m+n)\right) L_{-m}, \quad m \leqslant n.$$

Thus

$$\begin{aligned}
\mathrm{Ric}\left(\frac{L_n}{\sqrt{\theta_n}}, \frac{L_{-n}}{\sqrt{\theta_n}}\right) &= \sum_{m=n+1}^{\infty} \frac{2(m+n)\lambda_{m-n,n} - 2(m+2n)\lambda_{m,n}}{\theta_n} \\
&\quad - \sum_{m=1}^{n} \frac{(m+n)(2n-m) + 2(m+2n)\lambda_{m,n}}{\theta_n} \\
&= \sum_{m=1}^{n} \frac{2(m+2n)\lambda_{m,n}}{\theta_n} \\
&\quad - \sum_{m=1}^{n} \frac{(m+n)(2n-m) + 2(m+2n)\lambda_{m,n}}{\theta_n} \\
&= -\sum_{m=1}^{n} \frac{(m+n)(2n-m)}{\theta_n} = -\frac{13n^3 - n}{6\theta_n}.
\end{aligned}$$

Now we can use Proposition 3.1 to extend the result to all $n \in \mathbb{Z}$. □

Acknowledgments

The author thanks Ana Bela Cruzeiro and Jean-Claude Zambrini for organizing a satellite conference of the International Congress of Mathematicians on *Stochastic Analysis in Mathematical Physics* in September of 2006 at the University of Lisbon, Portugal, where the results of this paper have been presented.

References

1. H. Airault. Riemannian connections and curvatures on the universal Teichmuller space. *Comptes Rendus Mathematique* **341** (2005), 253–258.

2. H. Airault and V. Bogachev. Realization of Virasoro unitarizing measures on the set of Jordan curves. *C. R. Acad. Sci. Paris*, Ser. I, **336** (2003), 429–434.
3. H. Airault and P. Malliavin. Unitarizing probability measures for representations of Virasoro algebra. *J. Math. Pures Appl.* **80 (6)** (2001), 627–667.
4. H. Airault, P. Malliavin and A. Thalmaier. Support of Virasoro unitarizing measures. *C. R. Math. Acad. Sci. Paris*, Ser. I, **335** (2002), 621–626.
5. H. Airault, P. Malliavin and A. Thalmaier. Canonical Brownian motion on the space of univalent functions and resolution of Beltrami equations by a continuity method along stochastic flows. *J. Math. Pures Appl.* **83 (8)** (2004), 955–1018.
6. H. Airault and J. Ren. Modulus of continuity of the canonic Brownian motion "on the group of diffeomorphisms of the circle". *J. Funct. Anal.* **196** (2002), 395–426.
7. M. J. Bowick and S. G. Rajeev. The holomorphic geometry of closed bosonic string theory and $\mathrm{Diff}(S^1)/S^1$. *Nuclear Phys. B* **293** (1987), 348–384.
8. M. J. Bowick and S. G. Rajeev. String theory as the Kähler geometry of loop space. *Phys. Rev. Lett.* **58** (1987), 535–538.
9. B. Driver. A Cameron-Martin type quasi-invariance theorem for Brownian motion on a compact Riemannian manifold. *J. Funct. Anal.* **110 (2)** (1992), 272–376.
10. S. Fang. Canonical Brownian motion on the diffeomorphism group of the circle. *J. Funct. Anal.* **196** (2002), 162–179.
11. M. Gordina. Hilbert-Schmidt groups as infinite-dimensional Lie groups and their Riemannian geometry. *J. Func. Anal.* **227** (2005), 245–272.
12. M. Gordina and P. Lescot. Riemannian geometry of $\mathrm{Diff}(S^1)/S^1$. *J. Func. Anal.* **239** (2006), 611–630.
13. A. A. Kirillov. Geometric approach to discrete series of unirreps for Vir. *J. Math. Pures Appl.* **77** (1998), 735–746.
14. A. A. Kirillov and D. V. Yur'ev. Kähler geometry of the infinite-dimensional homogeneous space $M = \mathrm{Diff}_+(S^1)/\mathrm{Rot}(S^1)$, (Russian). *Funktsional. Anal. i Prilozhen.* **21** (1987), 35–46.
15. P. Malliavin. The canonic diffusion above the diffeomorphism group of the circle. *C.R. Acad. Sci. Paris* Sér. I Math. **329** (1999), 325–329.
16. K. Nomizu. Studies on Riemannian homogeneous spaces. *Nagoya Math. J.* **9** (1955), 43–56.
17. I. Shigekawa and S. Taniguchi. A Kähler metric on a based loop group and a covariant differentiation, in *Itô's stochastic calculus and probability theory*, 327–346, Springer, 1996.
18. S. Taniguchi. On almost complex structures on abstract Wiener spaces. *Osaka J. Math.* **33** (1996), 189–206.
19. B. Zumino. *The geometry of the Virasoro group for physicists*, in the NATO Advanced Studies Institute Summer School on Particle Physics, Cargese, France, 1987.

Ergodic theory of SDE's with degenerate noise

Antti Kupiainen*
*Department of Mathematics, Helsinki University,
P.O. Box 4, 00014 Helsinki, Finland
E-mail: ajkupiai@cc.helsinki.fi*

We review some topics in the study of stationary measures for stochastic differential equations with very degenerate noise. We sketch a rigorous derivation of Fourier's law from a system of closure equations for a nonequilibrium stationary state of a Hamiltonian system of coupled oscillators subjected to heat baths on the boundary.

Keywords: Hypoelliptic diffusions, non equilibrium stationary states, Fourier law

1. Stationary states for SDE's

Consider a stochastic differential equation

$$du(t) = X_0(u)dt + \sum_{a=1}^{n} X_a(u)db_a(t) \qquad (1)$$

where $u \in \mathbb{R}^N$, X_a are smooth vector fields and b_a i.i.d. Brownian motions. In this paper we are interested in the long time behaviour of the solutions to Eq. (1). Since the solution u gives rise to a Markov process (with suitable assumptions on the X_a) we need to understand the stationary states of this process. Thus, let $P_t(u_0, du)$ be the transition probabilities of the process, starting at u_0. An invariant (say Borel-) measure μ satisfies

$$\mu(A) = \int \mu(du_0) P_t(u_0, A) \qquad (2)$$

for Borel sets A. One would ideally like to prove that such μ exists, is unique and given any Borel measure μ_0 the measure

$$\mu_t(A) = \int \mu_0(du_0) P_t(u_0, A) \qquad (3)$$

---
*Research supported by the Academy of Finland

converges weakly to μ. Moreover, one would like to have estimates on the mixing rate i.e. the speed of the convergence. There are well known cases when such questions can be effectively answered.

One is the *elliptic* case i.e. when $\text{span}\{X_a(u)\}_{a=1}^n = N$ and the drift X_0 is not too strong. Then u is diffusive, μ is unique and depending on the drift the convergence is exponential or polynomial in time.

A more interesting case is the *hypoelliptic* one when

$$\text{span}\{X_a(u), \{X_\alpha, X_\beta\}, \{X_\alpha, \{X_\beta, X_\gamma\}\}, \ldots, a > 0, \alpha, \beta \cdots \geq 0\} = N. \quad (4)$$

Then, in good cases (say if the drift is sufficiently dissipative) μ is unique and mixing exponential. In physical terms, the drift spreads the degenerate noise to all degrees of freedom.

There are some interesting applications of the hypoelliptic case in non-equilibrium physics. Recently two cases have attracted a lot of attention of mathematical physicists, namely the case of two dimensional *turbulence* i.e. the two dimensional Navier-Stokes equations driven by a degenerate noise and the case of *heat conduction* in solids described below.

The incompressible two dimensional Navier-Stokes equation for the velocity field $u(t,x) \in \mathbb{R}^2$ may be written in terms of the vorticity field, $\omega = \partial_1 u_2 - \partial_2 u_1$. If we let the field be defined on the torus $x \in \mathbb{T} = (\mathbb{R}/2\pi\mathbb{Z})^2$ and go to Fourier transform $\omega_k(t) = \frac{1}{2\pi}\int_\mathbb{T} e^{ik\cdot x}\omega(t,x)dx$, with $k \in \mathbb{Z}^2$ the relation between u and ω becomes $u_k = i\frac{(-k_2,k_1)}{k^2}\omega_k$. The Navier-Stokes equations then become

$$d\omega(t) = X_0(\omega)dt + df(t), \quad (5)$$

where the drift is given by

$$X_0(\omega)_k = -\nu k^2 \omega_k + \frac{1}{2\pi}\sum_{l\in\mathbb{Z}^2\setminus\{0,k\}} \frac{k_1 l_2 - l_1 k_2}{|l|^2}\omega_{k-l}\omega_l \quad (6)$$

and df is (the curl of) the force acting on the fluid. $\nu > 0$ is the viscosity coefficient. (5) is of the form (1) with $N = \infty$ when df is suitably chosen. For the turbulence problem one is interested in forces involving only a finite number n of Fourier components:

$$df_k = \gamma_k (dx_k + i dy_k) \quad (7)$$

with $\gamma_k \neq 0$ for finitely many k and $x_k = x_{-k}$, $y_k = y_{-k}$ i.i.d Brownian motions. The system (5) satisfies the Hörmander condition (4), see Ref. 1. However, the infinite dimensionality of the problem makes the proof of ergodicity hard. Existence, uniqueness and exponential mixing of the stationary state were proven in Refs. 2–4 when sufficiently many of the γ_k are nonzero, depending on ν. For the harder problem, $n = O(1)$ independently on ν, existence and uniqueness were proven in Ref. 5.

However, not much more is known about the properties of the stationary state. On physical grounds one has a rather detailed picture of how the noise spreads in the system. One expects the state to exhibit fluxes of the conserved quantities of energy and enstrophy (i.e. mean square vorticity) as described by the Kolmogorov-Kraichnan theory (see Ref. 6 for a review). On the mathematical level it seems very hard to get hold of these phenomena due to the strongly interacting nature of the system. It would therefore be nice to have systems where the stationary state can be studied as a perturbation of a gaussian measure and nevertheless shown to exhibit a flux state.

One such problem concerns weakly nonlinear Hamiltonian systems coupled to boundary noise to which we turn now.

2. A model of heat conduction

One of the simplest and most fundamental phenomena in the physics of nonequilibrium systems is heat conduction in solids. A concrete physical example is a slab of crystal of linear extension N which is kept at the two ends at different temperatures T_1 and T_2. In this case, one would expect the crystal to exhibit a well defined temperature distribution along its length and also a constant flux of heat from the hot end to the cold one. Moreover these are believed to be described by a macroscopic equation, Fourier's law, which states that a local temperature gradient is associated with a flux of heat $\mathcal{J}$ which is proportional to the gradient:

$$\mathcal{J}(x) = -k(T(x))\nabla T(x) \qquad (8)$$

where the *heat conductivity* $k(T(x))$ is a function only of the temperature at x. For a constant flux in 1-direction, $\mathcal{J} = je_1$, T solves the ODE $(k(T)T')' = 0$ with boundary conditions T_1 and T_2 at 0 and N. Thus ∇T and $\mathcal{J}$ are $\mathcal{O}(1/N)$.

Despite its fundamental nature, a derivation of Fourier's law from first principles lies well beyond what can be mathematically proven (for reviews on the status of this problem, see Refs. 7–9). The quantities T and $\mathcal{J}$ in (8) are macroscopic variables, statistical averages of the variables describing the microscopic dynamics of matter. A first principle derivation of (8) entails a definition of T and $\mathcal{J}$ in terms of the microscopic variables and a proof of the law in some appropriate limit.

We will describe now some work (developed in Ref. 10) to make a step towards the proof of Fouriers's law in the context of a simple toy model of the crystal that fits (formally, since it is infinite dimensional) into the setup of Section 1. This system models atoms in the crystal lattice by coupled oscillators located in a strip V of width N in d-dimensional cubic lattice $\mathbb{Z}^d$. The oscillators are indexed by lattice points $x = (x_1, \ldots, x_d)$ with $0 \leq x_1 \leq N$ and the state of the oscillator at lattice point x is described by a momentum p_x and coordinate variable q_x, both of which for the sake of simplicity take values in $\mathbb{R}$ (instead of the more realistic $\mathbb{R}^d$). The dynamics of the oscillators consists of two parts: Hamiltonian dynamics in the bulk and noise on the boundary modelling heat baths at temperatures T_1 and T_2.

To define our model, consider a Hamiltonian function of the form

$$H(q,p) = \frac{1}{2}\sum_{x \in V} p_x^2 + \frac{1}{2}\sum q_x q_y \omega^2(x-y) + \frac{\lambda}{4}\sum_{x \in V} q_x^4 \qquad (9)$$

This describes a system of coupled anharmonic oscillators with coupling matrix ω^2 which we suppose to be short range. To describe the boundary noise, let

$$\gamma_x = \gamma(\delta_{x_1 0} + \delta_{x_1 N})$$

and

$$\tau_x = \sqrt{2\gamma T_1}\delta_{x_1 0} + \sqrt{2\gamma T_2}\delta_{x_1 N}$$

The dynamics is then given by

$$\begin{aligned} dq_x &= p_x dt \\ dp_x &= \left(-\frac{\partial H}{\partial q_x} - \gamma_x p_x\right) dt + \tau_x db_x \end{aligned} \qquad (10)$$

where b_x are i.i.d. Brownian motions.

The diffusion process (10) satisfies again the Hörmander condition (4), for example when $\omega^2 = -\Delta + r$ where Δ is the discrete Laplacean. When the boundary temperatures are equal, $T_1 = T_2 = T$ an explicit stationary state is given by the Gibbs state

$$Z^{-1} e^{-\beta H(q,p)} dq dp$$

of the Hamiltonian H with inverse temperature $\beta = 1/T$. This is the *equilibrium* situation. In the non-equilibrium case $T_1 \neq T_2$ there is no such simple formula and, indeed, in our setup, the existence of a stationary state is an open problem even in one dimension where the state space is finite dimensional (see Ref. 11 for a review of results in 1d first obtained in Ref. 12).

Supposing the existence of a stationary state, Fourier's law (8) is a statement about correlation functions in that state. One may define the local temperature as the expectation of the kinetic energy of the oscillator indexed by x:

$$T(x) = E\left(\frac{1}{2}p_x^2\right) \tag{11}$$

As for the heat flux, one writes the Hamiltonian H as a sum of local terms, each one pertaining to a single oscillator:

$$H = \sum_{x \in \Lambda} H_x.$$

Then, (10) leads to a local energy conservation law for x away from the boundaries:

$$\dot{H}_x + \sum_{\mu=1}^{d}(j_\mu(x+e_\mu) - j_\mu(x)) = 0,$$

where the current $j(x)$ is a quadratic function of p_y and q_y for y near x provided the coupling matrix ω^2 is short ranged. Then, the stationary heat current in eq. (8) is defined by

$$\mathcal{J}(x) = Ej(x).$$

The only rigorous results on deriving (8) for our model are for the harmonic case of a quadratic H.[13,14] In that case, Fourier's law *does not hold*: the current $j(x)$ is $\mathcal{O}(1)$ as $N \to \infty$ whereas $\nabla T = 0$ except near the boundary. If $\lambda \neq 0$ the law seems to hold in simulations in all dimensions.[15] It should be noted that the Hörmander condition holds for the harmonic case too, it is a result of the coupling matrix ω^2. Hence Fourier's law depends crucially on the nonlinear dynamics generated by the drift in eq. (10).

3. Closure equations

The nonequilibrium stationary state of our model may be studied via its correlation functions (the temperature and heat current being particular ones). Provided one has enough control of them, they will satisfy an infinite set of linear equations, the so called Hopf equations defined below, that are analogous to the BBGKY equations occuring in the kinetic theory of gases. We will outline a rigorous proof of Fourier's law[10] starting from a closure approximation to these equations. This is a nonlinear equation for the two point correlation functions obtained under the assumption that the stationary distribution be close to a Gaussian one. This, in turn, should hold when the strength λ of the anharmonic term in the Hamiltonian (9) is small. (For an earlier work on such a closure approximation, see Ref. 16).

We want to stress that what follows is not a complete proof of Fourier's law: presently we lack a rigorous justification of the closure approximation. However, the closure solution exhibits several features that one expects to hold more generally, such as nonlinear conductivity, and long range spatial correlations in the nonequilibrium state. As discussed below, the closure equation shares with the full model also the same kinetic scaling limit where the anharmonicity is taken to zero with N so that the mean free path is proportional to N. Such a scaling limit is governed by a Boltzmann equation describing phonon scattering.[9]

Thus let us suppose our SDE has a stationary state and expectations of polynomials in the q_x and p_x exist. We will study the state through such expectations, i.e. the correlation functions. Let us denote $(q_x, p_x) = (u_{1x}, u_{2x})$, $\Lambda(u)_{\alpha x} = -\lambda \delta_{\alpha,2} q_x^3$, $(\Gamma u)_x = (0, \gamma_x p_x)^T$. Then, (10) becomes

$$du(t) = \Big((A - \Gamma)u + \Lambda(u)\Big)dt + d\eta(t) \qquad (12)$$

where $A = \begin{pmatrix} 0 & 1 \\ -\omega^2 & 0 \end{pmatrix}$ and η denotes the noise terms. Define the stationary state correlation functions

$$G_n(x_1, \ldots x_n) = E(u_{x_1} \ldots u_{x_n})$$

(omitting indices, so that each u_{x_i} is either p_{x_i} or q_{x_i}). Then (10) and the Ito formula yield the Hopf equations

$$(A_n - \Gamma_n)G_n + \Lambda_n G_{n+2} + C_n G_{n-2} = 0, \qquad (13)$$

where $A_n = \sum A_{x_i}$ and Γ_n is defined similarly. Moreover,

$$\Lambda_n G_{n+2} = \sum_{i=1}^{n} E(u_{x_1} \ldots \Lambda(u)_{x_i} \ldots u_{x_n}),$$

$$C_n G_{n-2} = \sum_{i<j} C_{x_i x_j} G_{n-2}(x_1, ..., \hat{x}_i, ..., \hat{x}_j ... x_n)$$

where the arguments $\hat{x}_i, \hat{x}_j$ are missing. C equals one-half the time derivative of the covariance of η, i.e.

$$C = \begin{pmatrix} 0 & 0 \\ 0 & C \end{pmatrix},$$

with

$$C_{xy} = 2\gamma \delta_{xy}(T_1 \delta_{x_1 0} + T_2 \delta_{x_1 N}). \tag{14}$$

The equations (13) have the drawback that they do not "close": to solve for G_n, we need to know G_{n+2}.

We will now introduce an approximation that will lead to a closed set of nonlinear equations for G_2. For this we note that for λ small, the equilibrium $T_1 = T_2$ Gibbs state is close to Gaussian. When $T_1 \neq T_2$ we expect this to remain true and we look for a Gaussian approximation to equation (13) for small λ by means of a *closure*, i.e. replacing the G_n in terms of sums of products of G_2, using Wick's formula. If we simply take (13) for $n = 2$ and replace there G_4 by

$$\sum_p G_2(x_i, x_j) G_2(x_k, x_l)$$

where the sum runs over the pairings of $\{1, 2, 3, 4\}$ we obtain a quadratic equation for the covariance matrix G_2. It turns out that the solution to that equation is qualitatively similar to the $\lambda = 0$ case, i.e. G_2 does not exhibit a temperature profile nor a finite conductivity. The main effect of the nonlinearity turns out to be a renormalization of coupling matrix ω. Thus the closure has to be applied to the four point function equation. Indeed, similar closures were considered in the past in wave turbulence theory.[17]

Thus consider (13) for $n = 4$. The six point function may be written as

$$G_6 = \sum G_2 G_2 G_2 + \sum G_2 G_4^c + G_6^c \tag{15}$$

where G_4^c and G_6^c are the connected correlation functions (cumulants). Our approximation consists of dropping the last two terms on the RHS of (15). The first of these could be included in our analysis with minor modifications, but the main approximation is the dropping of the connected six point function. Both terms are of course subleading in the coupling constant λ.

Inserting this approximation to the $n = 4$ equation, solving G_4, and inserting the result into the $n = 2$ equation, leads to a nonlinear equation for G_2 of the form:

$$A_2 G_2 + \mathcal{N}(G_2) = \text{boundary terms}, \tag{16}$$

where the nonlinear term $\mathcal{N}(G_2)$ is given by

$$\mathcal{N}(G_2) = \Lambda_2 A_4^{-1} \Lambda_4 \sum G_2 G_2 G_2 \tag{17}$$

and the boundary terms involve Γ and C.

It is convenient to introduce the matrices

$$Q_{xy} = <q_x q_y>, \ P_{xy} = <p_x p_y>, \ J_{xy} = <q_x p_y>. \quad (18)$$

Clearly

$$\dot{Q} = J + J^T$$

so the (1,1)-component of (16) says $J_{xy} = -J_{yx}$ and we can write

$$G = \begin{pmatrix} Q & J \\ -J & P \end{pmatrix}.$$

The stationary state is translation invariant in the directions orthogonal to the 1-direction so we can write

$$G(x,y) = \int e^{ik(x-y)} \tilde{G}(x_1 + y_1, k) dk. \quad (19)$$

$\tilde{G}$ will be slowly varying in its first argument as $N \to \infty$.

Instead of writing $\mathcal{N}(G)$ explicitely (see Ref. 10 for that), we'll write $\mathcal{N}(G)$ in a *scaling limit*, the so called kinetic limit, where it takes a particularly simple form. In the kinetic limit one rescales x_1 to the unit interval and takes $N \to \infty$, $\lambda \to 0$, while keeping $R = N\lambda^2$ constant. Let

$$V(x,k) = \omega(k)\tilde{Q}(Nx,k) + iN^{-1}\tilde{J}(Nx,k)$$

for $x \in [0,1]$ and $k \in [0, 2\pi]^d$. Our equation (16) then becomes the following equation as $N \to \infty$

$$\nabla_k \omega(k) \nabla_x V(x,k) - RN(V) = \text{boundary terms} \quad (20)$$

with

$$N(V)(x,k) = \frac{9\pi^2}{2} \int dk_1 dk_3 dk_3 (\omega(k)\omega(k_1)\omega(k_2)\omega(k_3))^{-1}.$$
$$\delta(\omega(k) + \omega(k_1) - \omega(k_2) - \omega(k_3))\delta(k + k_1 - k_2 - k_3).$$
$$[V_1 V_2 V_3 - V(V_1 V_2 + V_1 V_3 - V_3 V_3)]. \quad (21)$$

where $V_i = V(x, k_i)$ and $V = V(x,k)$. (20) is a Boltzmann equation for a gas of *phonons*. (see Ref. 9 for a discussion of the kinetic theory of phonon systems). The collision term (21) describes scattering of phonons with momentum k and energy $\omega(k)$.

From perturbation theory one expects the full Hopf equations to reduce, in the kinetic limit, to the equations (20), (21). Thus our closure equations should have the same kinetic scaling limit as the full theory. It should be emphasized, however, that they are more general than the kinetic equations.

Indeed, we do not take any limit: λ, τ and N are fixed and we prove that our solution has the expected properties, with precise bounds on the remainder, for λ small and N large. Moreover our equations have *long range correlations* that are absent in the Boltzmann equation.

We will now explain the argument for solving (16) and (21). An important property of $\mathcal{N}$ and N is that, for all T, $\mathcal{N}(TV^{(0)}) = 0$ where $V^{(0)} = \omega^{-1}$. For $\gamma = 0$ these form a 1-parameter family of solutions of (16) and (21). In the equilibrium case $T_1 = T_2$ and $\gamma \neq 0$ only one of these persists, namely the one with $T = T_1$. This is the analogue for the closure equation of the true equilibrium Gibbs state which has $V = V^{(0)} + \mathcal{O}(\lambda)$.

Since we are looking for a solution that is locally in x_1 close to this equilibrium we need to understand the linearization of the nonlinear term $\mathcal{N}$ at the constant temperature equilibrium $TV^{(0)}$. For (21) this is the linearized Boltzmann operator

$$N(TV^{(0)} + v)(x, k) = \mathcal{L}(v(x,\cdot)(k) \tag{22}$$

where $\mathcal{L}$ is a sum of a multiplication and an integral operator

$$(\mathcal{L}f)(k) = A(k)f(k) + \int B(k, k')f(k')dk'. \tag{23}$$

For (16) the linearization acts also on the slowly varying coordinate x, however diagonally in momentum space:

$$(\mathcal{L}\hat{v})(p, k) = A(p, k)\hat{v}(p, k) + \int B(p, k, k')\hat{v}(p, k')dk'. \tag{24}$$

The integral kernel $B(p, k, k')$ is of the form

$$B(p, k, k') = = \sum_{\mathbf{s}} \int \Delta\Big(\sum_{i=1}^{2} s_i \omega(k_i) + s_3 \omega(k' + p) + s_4 \omega(p - k)\Big)$$
$$\cdot \delta(k - k_1 - k_2 - k')\rho_{\mathbf{s}} dk_1 dk_2 \tag{25}$$

where $\Delta(x) = \delta(x)$ or $\mathcal{P}\left(\frac{1}{x}\right)$ and $\rho_{\mathbf{s}}$ is a combination of ω's. $A(p, k)$ is given by a similar expression integrated over k'. At $p = 0$ these operators reduce to the linearized Boltzmann operators (23), but, in general, they mix the slow and the fast coordinates.

Since we are looking for solutions that are slowly varying in x i.e. sharply peaked in momentum space around $p = 0$, it is important to understand (24) at $p = 0$ i.e the operator (23). This operator has *two zero modes*. One of them is easy to understand. Since $N(TV^{(0)}) = 0$ for all T, taking derivative with respect to T, one finds $\mathcal{L}\omega^{-1} = 0$. There is, however, also a second zero mode: $\mathcal{L}\omega^{-2} = 0$.

While the first zero mode has to persist for the full Hopf equations due to the one parameter family of Gibbs states that solve them for $\gamma = 0$, the second one is an artifact of the closure approximation. The phonon scattering described by the nonlinear term conserves phonon energy, leading to the first zero mode, and also phonon number, leading to the second one. The connected six-point correlation function which was neglected in the closure approximation would produce terms that violate phonon number conservation and remove the second zero mode. However, for weak anharmonicity, its eigenvalue would be close to zero and should be treated as some perturbation of the present analysis.

The second zero mode leads one to expect that our equations have, in the $\gamma = 0$ limit, a two parameter family of stationary solutions, which is indeed the case. These are given by

$$V_{T,A}(x,k) = T(\omega(k) - A)^{-1}. \tag{26}$$

The second zero mode is proportional to the derivative of $V_{T,A}$ with respect to A, at $A = 0$. We are then led to look for solutions in the form

$$V(x,k) = V_{T(x),A(x)}(k) + v(x,k), \tag{27}$$

where the first term is of *local equilibrium* form with slowly varying temperature and "chemical potential" profiles $T(x)$ and $A(x)$ and where v is a perturbation orthogonal to the zero modes in a suitable inner product.

Let us now describe our main results, both concerning (16) and (21) (for (21), see Ref. 18). Apart from boundary effects, $V(x,k)$ takes the form:

$$V(x,k) = V_{T(x),A(x)}(k) + o\left(\frac{1}{N}\right), \tag{28}$$

where $A(x) = \mathcal{O}(\tau\lambda^2)$ and $T(x) = T_1 + \frac{|x|}{N}(\tau + \mathcal{O}(\tau^2 + \tau\lambda^2))$.

Corresponding to the two conservation laws in this model, we have two currents (j, j') that are related to the discrete gradients $\nabla T, \nabla A$, for x not too close to the boundaries, by:

$$\begin{pmatrix} j(x) \\ j'(x) \end{pmatrix} = -\kappa(A(x)) \begin{pmatrix} T^{-2}\nabla T(x) \\ T^{-1}\nabla A(x) \end{pmatrix} + o\left(\frac{1}{N}\right) \tag{29}$$

where $\kappa(A(x))$ is a 2×2 matrix whose matrix elements are analytic functions of $A(x)$ and are $\mathcal{O}(\lambda^{-2})$. Moreover, for $x \neq 0, N$,

$$j(x+1) - j(x) = 0, \tag{30}$$

and, for x not too close to the boundaries,

$$j'(x+1) - j'(x) = f(T(x), A(x), \nabla T(x), \nabla A(x)) + o\left(\frac{1}{N^2}\right), \tag{31}$$

where f is analytic in its arguments and is quadratic in $\nabla T(x)$, $\nabla A(x)$, i.e. of order $\frac{1}{N^2}$. The equations (29) and (31) can be viewed as a (discrete) nonlinear elliptic system for the leading terms of T and A. However we don't solve them directly but rather derive a coupled system of equations for T, A and the rest of G which we solve by fixed point methods in a suitable space.

The temperature profile $T(x)$ is *non-linear*. Due to the complicated functions κ and f in equations (29) and (31) there is no closed formula. However, in the *kinetic limit* $\lambda = \mathcal{O}(1/\sqrt{N})$ these formulae simplify. A will disappear and the *inverse temperature* has a linear profile:

$$T(x)^{-1} = T_2^{-1} + \frac{|x|}{N}(T_1^{-1} - T_2^{-1})$$

which results from the Fourier law:

$$j(x) = -\kappa\, T(x)^{-2}\nabla T(x) + o\left(\frac{1}{N}\right).$$

The main technical assumptions we need to prove these claims are the smallness of $\frac{1}{N}$, λ and ΔT. For convenience, we take also the coupling γ to the reservoirs small, as $\gamma = N^{-1+\alpha}$ for some small $\alpha > 0$. The coupling γ is important to fix the boundary conditions for the elliptic equations determining T and A, but, for this, one only needs it to be bigger than N^{-1}. We also need for our analysis that ω^2 is sufficiently *pinning*; in fact, we choose $\omega^2 = (-\Delta + m^2)^2$, with m large enough, i.e. in momentum space $\omega(k) = 2\sum_{i=1}^{d}(1 - \cos k_i) + m^2$. This choice simplifies several estimates.

The most crucial assumption is that the space dimension has to be at least three (or at least two for the Boltzmann equation (21), see Ref. 18). This is because of the low regularity of the collision kernels in eq. (25). To prove the spectral properties of the linear operators (23), we need compactness of the integral operator B in a nice enough space. In three dimensions, this holds in a space of Hölder continuous functions whereas in two dimensions this is not the case. Even in three dimensions, the resulting solutions have low regularity in momentum space which translates into long range correlations in physical space.

4. Conclusions

The study of the stationary states of degenerate hypoelliptic diffusions is still in its infancy. The physically interesting questions like the existence of fluxes of conserved quantities require a much more detailed understanding of the nonlinear dynamics that has been achieved so far. Noise should help with some problems like wiping out elliptic regions (eg. KAM tori in the Hamiltonian systems), but transport is a result of the deterministic dynamics the noise setting only boundary conditions. This is evident in the system discussed above where the macroscopic equation was the product of the phonon scattering process and noise determined its boundary conditions.

Systems with weak nonlinearites such as the one discussed above are promising candidates for rigorous study since the non-equilibrium measure should be close to a gaussian one. However, this gaussian measure is a singular perturbation in the coupling constant of the harmonic model. If λ^2 is too small compared with $1/N$ the system behaves as the harmonic one with no temperature profile and $\mathcal{O}(1)$ current. The first step in our model beyond the closure should be the study of the kinetic limit. The derivation of the Boltzman equation in that limit has been achieved in a harmonic model where the masses of the oscillators were taken random.[19] It would be very interesting to extend that analysis to the model described in this paper.

References

1. W. E, and J. Mattingly. Ergodicity for the Navier-Stokes equation with degenerate random forcing: finite-dimensional approximation. *Comm. Pure Appl. Math.* **54(11)** (2001), 1386–1402.
2. J. Bricmont, A. Kupiainen, and R. Lefevere. Exponential mixing of the 2D stochastic Navier-Stokes dynamics. *Comm. Math. Phys.* **230(1)** (2002), 87–132.
3. W. E, J. Mattingly, and Ya G. Sinai. Gibbsian dynamics and ergodicity for the stochastic forced Navier-Stokes equation. *Comm. Math. Phys.* **224(1)** (2001).
4. S. Kuksin, and A. Shirikyan. Coupling approach to white-forced nonlinear PDEs. *J. Math. Pures Appl. (9)* **81(6)** (2002), 567–602.
5. M. Hairer, and J. Mattingly. Ergodicity of the degenerate stochstic 2D Navier-Stokes equation. *Annals of Mathematics* **164(3)** (2006).
6. D. Bernard. Turbulence for (and by) amateurs. http://arxiv.org/pdf/cond-mat/0007106
7. F. Bonetto, J. L. Lebowitz, and L. Rey-Bellet. Fourier Law: A challenge to Theorists. In: *Mathematical Physics 2000*, Imp. Coll. Press, London 2000, 128–150.
8. S. Lepri, R. Livi, and A. Politi. Thermal conductivity in classical low-dimensional lattices. *Physics Reports* **377** (2003), 1–80.
9. H. Spohn. The phonon Boltzmann equation, properties and link to weakly anharmonic lattice dynamics. *J. Stat. Phys.*. To appear, arXiv:math-phys/0505025.

10. J. Bricmont, and A. Kupiainen. Towards a derivation of Fourier's law for coupled anharmonic oscillators. To appear in *Comm. Math. Phys.*.
11. L. Rey-Bellet. Nonequilibrium statistical mechanics of open classical systems. In:*XIVTH International Congress on Mathematical Physics*, edited by Jean-Claude Zambrini, World Scientific, Singapore, 2006.
12. J.-P. Eckmann, C.-A. Pillet, and L. Rey-Bellet. Non-equilibrium statistical mechanics of anharmonic chains coupled to two heat baths at different temperatures. *Comm. Math. Phys.* **201** (1999), 657–697.
13. H. Spohn, and J. L. Lebowitz. Stationary non-equilibrium states of infinite harmonic systems. *Comm. Math. Phys.* **54** (1977), 97–120.
14. Z. Rieder., J. L. Lebowitz, and E. Lieb. Properties of a harmonic crystal in a stationary non-equilibrium state. *J. Math. Phys.* **8** (1967), 1073–1085.
15. K. Aoki, and D. Kusnezov. Nonequilibrium statistical mechanics of classical lattice ϕ^4 field theory. *Ann. Phys.* **295** (2002), 50–80.
16. R. Lefevere, and A. Schenkel. Normal heat conductivity in a strongly pinned chain of anharmonic oscillators. *J. Stat. Mech.* L02001 (2006). Available on: http://www.iop.org/EJ/toc/1742-5468/2006/02.
17. V. E. Zakharov, V. S. Lvov, and G. Falkovich. Kolmogorov spectra of turbulence I: wave turbulence. Springer, Berlin (1992).
18. J. Bricmont, and A. Kupiainen. Approach to equilibrium for the phonon Boltzmann equation. Preprint.
19. J. Lukkarinen, and H. Spohn. Kinetic Limit for Wave Propagation in a Random Medium. http://arxiv.org/abs/math-ph/0505075

Dynkin's isomorphism without symmetry

Yves Le Jan

Mathématiques
Université Paris 11
91405 Orsay, France
E-mail: yves.lejan@math.u-psud.fr

We extend Dynkin isomorphism for functionals of the occupation field of a symmetric Markov processes and of the associated Gaussian field to a suitable class of non symmetric Markov processes.

Keywords: Markov processes, Gaussian processes

1. Introduction

The purpose of this note is to extend Dynkin isomorphism involving functionals of the occupation field of a symmetric Markov processes with non polar points and of the associated Gaussian field to a suitable class of non symmetric Markov processes. This was briefly proposed in Ref. 5 using Grassmann variables, extending to the non symmetric case some results of Ref. 4. Here we propose an alternative approach, not relying on Grassmann variables that can be applied to the study of local times, in the spirit of Ref. 6. It works in general on a finite space and on an infinite space under some assumption on the skew symmetric part of the generator which is checked on two examples.

2. The finite case

In this section, we will prove two formulas given in Proposition 1 and Corollary 2 which relate the local time field of a non symmetric Markov process on a finite space to the square of the associated complex "twisted" Gaussian field. The result and the proof appear to be a direct extension of Dynkin's isomorphism.

2.1. Dual processes

Let us first consider the case of an irreducible Markov process on a finite space X, with finite lifetime ζ, generator L and potential $V = (-L)^{-1}$.

Let $m = \mu V$ for any nonnegative probability μ on X. Recall that L can be written in the form $L = q(I - \Pi)$ with q positive and Π submarkovian. Then the m-adjoint $\widehat{L}$ can be expressed similarly with the same q and a possibly different submarkovian matrix $\widehat{\Pi}$. Moreover, $m = \widehat{\mu}\widehat{V}$, with $\widehat{\mu}$ the law of $x_{\zeta-}$ under $\mathbb{P}_\mu$.

Note that for any $z = x + iy \in \mathbb{C}^X$, the "energy" $Re(\langle Lz, \overline{z}\rangle_m) = Re(\sum -(Lz)_u \overline{z}_u m_u)$ is nonnegative as it can be written $\frac{1}{2}\left(\sum C_{u,v}(z_u - z_v)(\overline{z_u} - \overline{z_v}) + \left\langle \Pi 1 + \widehat{\Pi} 1 - 2, z\overline{z}\right\rangle_m\right)$, with $C_{u,v} = C_{v,u} = m_u q_u \Pi_{u,v}$. The highest eigenvector of $\frac{1}{2}(\Pi + \widehat{\Pi})$ is nonnegative by the well known argument which shows that the module contraction lowers the energy and it follows from the strict submarkovianity that the corresponding eigenvalue is strictly smaller than 1. Hence there is a "mass gap": For some positive ε, the "energy" $Re(\langle -Lz, \overline{z}\rangle_m)$ dominates $\varepsilon \langle z, \overline{z}\rangle_m$ for all z.

2.2. A twisted Gaussian measure

Then, although L is not symmetric, an elementary computation (given in a more general context in the following section) shows that for any $\chi \in \mathbb{R}_+^X$, denoting M_χ the diagonal matrix with coefficients given by χ,

$$\frac{1}{(2\pi)^{|X|}} \int (e^{-\langle z\overline{z}, \chi\rangle_m} e^{\frac{1}{2}\langle Lz, \overline{z}\rangle_m} \Pi dx_u dy_u = \det(-M_m L + M_{\chi m})^{-1}.$$

As a consequence, differentiating with respect to χ_x,

$$\frac{1}{(2\pi)^{|X|}} \int z_x \overline{z_x} (e^{-\langle z\overline{z}, \chi\rangle_m} e^{\frac{1}{2}\langle Lz, \overline{z}\rangle_m} \Pi dx_u dy_u =$$

$$\det(-M_m L + M_{\chi m})^{-1} \frac{1}{m_x}(-L + M_\chi)^{-1}_{xx}$$

In a similar way using perturbation by a non diagonal matrix, one obtains

$$\frac{1}{(2\pi)^{|X|}} \int z_x \overline{z_y} (e^{-\langle z\overline{z}, \chi\rangle_m} e^{\frac{1}{2}\langle Lz, \overline{z}\rangle_m} \Pi dx_u dy_u =$$

$$\det(-M_m L + M_{\chi m})^{-1} \frac{1}{m_y}(-L + M_\chi)^{-1}_{xy}$$

But with the usual notations for Markov processes, setting $l_t^x = \int_0^{t\wedge\zeta} 1_{\{x_s=x\}} \frac{1}{m_{x_s}} ds$ and $l_\zeta^x = l^x$, we have

$$\frac{1}{m_y}(-L + M_\chi)_{xy}^{-1} = \mathbb{E}_x\left(\int_0^\zeta e^{-\langle \chi, l_t\rangle_m} dl_t^y\right)$$

Defining the path measure $\mathbb{E}_{x,y}$ by: $\mathbb{E}_x(\int_0^\zeta G(x_s, s \leq t) dl_t^y) = \mathbb{E}_{x,y}(G)$ the above relation writes

$$\frac{1}{m_y}(-L + M_\chi)_{xy}^{-1} = \mathbb{E}_{x,y}(e^{-\langle \chi, l\rangle_m})$$

It follows that we have proved the following:

Proposition 2.1. *For any continuous function F on $\mathbb{R}_+^X$*

$$(*) \quad \int z_x \bar{z}_y F(z_u \bar{z}_u, u \in X) e^{\frac{1}{2}\langle Lz, \bar{z}\rangle_m} \Pi dx_u dy_u =$$

$$\int \mathbb{E}_{x,y}(F(l^u + z_u \bar{z}_u, u \in X)) e^{\frac{1}{2}\langle Lz, \bar{z}\rangle_m} \Pi dx_u dy_u$$

2.3. *Positivity*

It should be noted that setting $\rho_u = \frac{1}{2} z_u \bar{z}_u$ and $z_x = \sqrt{\rho_x/2} e^{i\theta_x}$ the image on $\mathbb{R}_+^X$ of the normalized complex measure $\nu_X = \frac{1}{(2\pi)^{|X|}} \det(-M_m L) e^{\frac{1}{2}\langle Lz, \bar{z}\rangle_m} \Pi dx_u dy_u$ by the map $z_u \to \rho_u$ is an infinitely divisible probability distribution Q on $\mathbb{R}_+^X$ with density $\frac{1}{(2\pi)^{|X|}} \det(-M_m L) \int e^{\langle L\sqrt{\rho}e^{i\theta}, \sqrt{\rho}e^{-i\theta}\rangle_m} \Pi d\theta_u$. Note that the positivity is not a priori obvious when L is not m-symmetric. This important fact follows easily by considering the moment generating function $\Phi(s) = \frac{\det(-L)}{\det(-L+M_s)} = \det(I + (-L)^{-1} M_s)^{-1}$ defined for all s with non negative coordinates, positive and analytic. The expansion in power series around any s (which appears for example in Ref. 8) is explicit:

$$\frac{\Phi(s+h)}{\Phi(s)} = \det(I + (-L + M_s)^{-1} M_h)^{-1}$$

$$= \exp(-\log(\det(I + (-L + M_s)^{-1} M_h)))$$

$$= \exp\left(\sum \frac{(-1)^k}{k} Tr([(-L + M_s)^{-1} M_h]^k)\right).$$

As $(-L+M_s)^{-1}$ is nonnegative, it implies that Φ is completely monotone as in this last expression, all coefficients of h-monomials of order n are of the sign of $(-1)^n$. [a]

---

[a]Completely monotone functions in several variables were already used in Ref. 1.

Note that the same argument works for fractional powers of $\Phi(s)$ which shows the infinite divisibility. Let us incidentally mention it has been known for a long time (cf. Ref. 8) that this expansion can be simplified further in terms of permanents.

For $x = y$, the above proposition then yields the following:

Corollary 2.1. *For any continuous function F on $\mathbb{R}_+^X$*

$$(**) \quad \int \rho_x F(\rho_u, u \in X) Q(d\rho) = \int \mathbb{E}_{x,x}(F(l^u + \rho_u, u \in X)) Q(d\rho)$$

Note that this last formula is proved in Ref. 3 from a direct definition of the measure Q.

Another interpretation of this positivity and of infinite divisibility can be given in terms of a Poisson process of loops. It will be developed in a forthcoming paper but let us simply mention that Q appears to be equal to the distribution of the occupation field associated with the Poisson process of loops canonically defined by the Markov chain.

Remark 2.1. If $Y \subset X$, it is well known that the trace of the process on Y is a Markov process the potential of which is the restriction of V to $Y \times Y$. The distribution ν_X induces ν_Y. Therefore the formulas (*) and (**) on X and Y are consistent.

Example 2.1. Let us consider, as an example, the case where $X = \{1, 2....N\}$, $q_i = 1$, $\Pi_{i,j} = \mathbf{1}_{i<N, j=i+1}$, $\mu_i = \mathbf{1}_{i=1}$, $m_i = 1$ and $(-L)^{-1}_{i,j} = \mathbf{1}_{i \leq j}$.

The characteristic polynomial of $\frac{1}{2}(\Pi + \widehat{\Pi})$ is $(-\lambda + \sqrt{\lambda^2 - 1})^N + (-\lambda - \sqrt{\lambda^2 - 1})^N$ hence one gets easily that the mass gap equals $2 \sin^2\left(\frac{\pi}{2N}\right)$.

Under $\mathbb{E}_{x,x}$ all local times vanish except l^x which follows an exponential distribution. Moreover an easy calculation shows that Q reduces to a product of exponential distributions. The formula (**) reduces to the convolution of two exponentials.

3. The infinite case

We now explain how in certain situations, the above can be extended to a Markov process on an infinite space X. Of course, a Markov process for which points are not polar can always be viewed elementarily as a consistent system of processes on finite subspaces but we aim at a stronger representation allowing to consider any functional of the occupation field. There are some obvious obstructions to a generalization. The mass gap property does not always hold: consider for example the case of a constant drift on an interval, analogous to the above example. Some assumptions have to be made in order that the energy controls the skew-symmetric part of the generator.

3.1. Some calculations in Gaussian space

Let H be a real Hilbert space with scalar product $<,>$. At first the reader may suppose it finite dimensional and then check that the assumptions we will make allow to extend the results to the infinite dimensional case.

Let ϕ be the canonical Gaussian field indexed by H. Given any ONB e_k of H, $w_k = \phi(e_k)$ are independent normal variables. Recall that for all $f \in H$, $\phi(f) = \sum_k <f, e_k> w_k$ and $E(e^{i\phi(f)}) = e^{-\frac{\|f\|^2}{2}}$.

In the following, $\phi(f)$ can be denoted by $\langle \phi, f \rangle$ though of course ϕ does not belong to H in general.

Let K be any Hilbert-Schmidt operator on H. Note that $K\phi = \sum_k w_k K e_k$ is well defined as a H-valued random variable, and that $E(\|K\phi\|^2) = Tr(KK^*)$.

Let C be a symmetric non negative trace-class linear operator on H. Recall that the positive integrable random variable $\langle C\phi, \phi \rangle \in \mathbb{L}^1$ can be defined by $\sum_k \langle Ce_k, e_k \rangle w_k^2$ for any ONB diagonalizing C.

Moreover, $E\left(e^{-\frac{1}{2}\langle C\phi,\phi\rangle+i\phi(f)}\right) = \det(I+C)^{-\frac{1}{2}} e^{-\frac{\langle (I+C)^{-1}f, f\rangle}{2}}$ (the determinant can be defined as $\prod(1+\lambda_i)$, where the λ_i are the eigenvalues of C.

In fact $\det(I+T)$ is well defined for any trace class operator T as $1 + \sum_{n=1}^{\infty} Tr(T^{\wedge n})$ (Cf. Ref. 7 Chapter 3). It extends continuously the determinant defined with finite ranks operators and it verifies the identity: $\det(I + T_1 + T_2 + T_1 T_2) = \det(I + T_1)\det(I + T_2)$. By Lidskii's theorem, it is also given by the product $\prod(1 + \lambda_i)$ defined by the eigenvalues of the trace class (hence compact) operator T, counted with their algebraic multiplicity.

Let ϕ_1 and ϕ_2 be two independent copies of the canonical Gaussian process indexed by H. Let B be a skew-symmetric Hilbert-Schmidt operator on H. Note that $\langle B\phi_1, \phi_2 \rangle = \sum \langle Be_k, e_l \rangle w_k^1 w_l^2 = -\langle B\phi_2, \phi_1 \rangle$ is well defined in $\mathbb{L}^2$ and that $E\left(e^{i\langle B\phi_1, \phi_2\rangle}\right) = E\left(e^{-\frac{1}{2}\|B\phi_1\|^2}\right) = \det(I + BB^*)^{-\frac{1}{2}}$.

As B is Hilbert-Schmidt, BB^* is trace-class. The renormalized determinant $\det_2(I + B) = \det((I + B)e^{-B})$ is well defined (Cf. Ref. 7), and as the eigenvalues of B are purely imaginary and pairwise conjugated, it is strictly positive. Moreover since $B^* = -B$, $\det_2(I+B) = \det_2(I-B) = $

$\det(I + BB^*)^{\frac{1}{2}}$.

Finally, it comes that $E\left(e^{i\langle B\phi_1,\phi_2\rangle}\right) = \det_2(I+B)^{-1}$.

More generally, setting $\psi = \phi_1 + i\phi_2$,
$$E\left(e^{-\frac{1}{2}\langle (C-B)\psi,\overline{\psi}\rangle}\right) = \det_2(I+C+B)^{-1}\exp(-Tr(C))$$
(Recall that when T is trace class, $\det_2(I+T)\exp(Tr(T)) = \det(I+T)$).

Indeed, $E\left(e^{-\frac{1}{2}\langle (C-B)\psi,\overline{\psi}\rangle}\right) = E\left(e^{-\frac{1}{2}\langle C\phi_1,\phi_1\rangle - \frac{1}{2}\langle C\phi_2,\phi_2\rangle + i\langle B\phi_1,\phi_2\rangle}\right)$

$= \det(I+C)^{-\frac{1}{2}} E\left(e^{-\frac{1}{2}(\langle (I+C)^{-1}B\phi_1,B\phi_1\rangle + \langle C\phi_1,\phi_1\rangle)}\right)$ (by integration in ϕ_2)

$= \det(I+C)^{-\frac{1}{2}} \det(I+C - B(I+C)^{-1}B)^{-\frac{1}{2}}$

$= \det\left(I - (I+C)^{-\frac{1}{2}}B(I+C)^{-1}B(I+C)^{-\frac{1}{2}}\right)^{-\frac{1}{2}} \det(I+C)^{-1}$

$= \det\left(\left(I + (I+C)^{-\frac{1}{2}}B(I+C)^{-\frac{1}{2}}\right)\left(I - (I+C)^{-\frac{1}{2}}B(I+C)^{-\frac{1}{2}}\right)\right)^{-\frac{1}{2}} \det(I+C)^{-1}$

$= \det_2\left(I + (I+C)^{-\frac{1}{2}}B(I+C)^{-\frac{1}{2}}\right)^{-1} \det(I+C)^{-1}$ (since $\det_2\left(I + (I+C)^{-\frac{1}{2}}B(I+C)^{-\frac{1}{2}}\right) = \det_2\left(I - (I+C)^{-\frac{1}{2}}B(I+C)^{-\frac{1}{2}}\right)$ by skew symmetry as before)

$= \det_2(I+C+B)^{-1}\exp(-Tr(C))$.

Note that $I+C+B$ is always invertible, as $C+B$ is a compact operator and -1 is not an eigenvalue.

Let f_1 and f_2 be two elements of H. Set $D(f) = \langle f,f_1\rangle f_2$. For small enough ε, $E\left(e^{-\frac{1}{2}(\langle (C-B)\psi,\overline{\psi}\rangle + \varepsilon\psi(f_1)\overline{\psi}(f_2))}\right)$
$= \det_2(I+C+B+\varepsilon D)^{-1}\exp(-Tr(C+\varepsilon D))$
$= \det_2(I+C+B)^{-1}\exp(-Tr(C))\det(I+\varepsilon(I+C+B)^{-1}D)^{-1}$.

Hence, differentiating both members at $\varepsilon = 0$,

$E\left(\psi(f_1)\overline{\psi}(f_2)e^{-\frac{1}{2}(\langle (C-B)\psi,\overline{\psi}\rangle)}\right) =$
$$\det_2(I+C+B)^{-1}\exp(-Tr(C))Tr((I+C+B)^{-1}D)$$

Therefore

$$\frac{E\left(\psi(f_1)\overline{\psi}(f_2)e^{-\frac{1}{2}(\langle (C-B)\psi,\overline{\psi}\rangle)}\right)}{E\left(e^{-\frac{1}{2}(\langle (C-B)\psi,\overline{\psi}\rangle)}\right)} = \langle (I+C+B)^{-1}(f_1), f_2\rangle.$$

If C is only Hilbert-Schmidt, we can consider only the renormalized "Wick square" : $\langle C\phi, \phi \rangle := \sum_k \langle Ce_k, e_k \rangle (w_k^2 - 1)$ for any ONB diagonalizing C and $E\left(e^{-\frac{1}{2}:\langle C\phi,\phi\rangle:+i\phi(f)}\right) = \det_2(I+C)^{-\frac{1}{2}} e^{-\frac{\langle (I+C)^{-1}f,f\rangle}{2}}$

The results given above extend immediately as follows:

$$E\left(e^{-\frac{1}{2}:\langle (C-B)\psi,\overline{\psi}\rangle:}\right) = \det_2(I+C+B)^{-1}$$

$$\frac{E\left(\psi(f_1)\overline{\psi}(f_2) e^{-\frac{1}{2}\left(:\langle (C-B)\psi,\overline{\psi}\rangle:\right)}\right)}{E\left(e^{-\frac{1}{2}\left(:\langle (C-B)\psi,\overline{\psi}\rangle:\right)}\right)} = <(I+C+B)^{-1}(f_1), f_2>.$$

3.2. A class of Markov processes in duality

Let $(V_\alpha, \alpha \geq 0)$ and $\widehat{V}_\alpha$ be two Markovian or submarkovian resolvents in duality in a space $\mathbb{L}^2(X, \mathcal{B}, m)$ with generators L and $\widehat{L}$, such that:

1) Denoting $\mathcal{D} = \mathcal{D}(L) \cap \mathcal{D}(\widehat{L})$, $L(\mathcal{D})$ is dense in $\mathbb{L}^2(m)$
2) $< -Lf, f>_m \geq \varepsilon < f, f>_m$ for some $\varepsilon > 0$ and any $f \in \mathcal{D}$ (i.e. we assume the existence of a spectral gap: it can always be obtained by adding a negative constant to L.

Let H be the completion of $\mathcal{D}$ with respect to the energy norm. It is a functional space space imbedded in $\mathbb{L}^2(X, \mathcal{B}, m)$. Let A be the associated self adjoint generator so that $H = \mathcal{D}(\sqrt{-A})$. On $\mathcal{D}$, $\frac{1}{2}(L + \widehat{L}) = A$.

The final assumption is crucial to allow the control of the antisymmetric part:

3) $B = (-A)^{-1} \frac{L - \widehat{L}}{2}$ is a Hilbert-Schmidt operator on H.

Equivalently, $(-A)^{-\frac{1}{2}} \frac{L - \widehat{L}}{2} (-A)^{-\frac{1}{2}}$ is an antisymmetric Hilbert Schmidt operator on $\mathbb{L}^2(m)$ since for any ONB e_k of H, $(-A)^{\frac{1}{2}} e_k$ is an ONB of $\mathbb{L}^2(m)$. Note that $I - B$ is bounded and invertible on H and that $V_0 = (I - B)^{-1}(-A)^{-1}$ maps $\mathbb{L}^2(m)$ into H. Indeed, one can see first that on $\mathcal{D}$, $A(I - B) = L$ so that on $L(\mathcal{D})$, $V_0 = (I - B)^{-1}(-A)^{-1}$

This applies to the case of the finite space considered above.
Let us mention other examples:

Example 3.1. Diffusion with drift on the circle: $X = S^1$, $A = \frac{\partial^2}{\partial \theta^2} - \varepsilon$, $L - \widehat{L} = b(\theta) \frac{\partial}{\partial \theta}$, where b is a bounded function on S^1.

Indeed, considering the orthonormal basis $e^{ik\theta}$ in $\mathbb{L}^2(d\theta)$, $\frac{1}{\sqrt{k^2+\varepsilon}}e^{ik\theta}$ is an orthonormal basis in $H = H^1$, and

$$\sum_k \left\| (-A)^{-\frac{1}{2}} b(\theta) \frac{\partial}{\partial \theta} \frac{1}{\sqrt{k^2+\varepsilon}} e^{ik\theta} \right\|_{\mathbb{L}^2(d\theta)}^2 = \sum_{k,l} \frac{k^2}{k^2+\varepsilon} (\widehat{b}(l-k))^2 \frac{1}{l^2+\varepsilon} < \infty$$

Example 3.2. Levy processes on the circle: The Fourier coefficients $a_k + ib_k$ of L should verify $\sum_k \left(\frac{b_k}{a_k}\right)^2 < \infty$.

3.3. An extension of Dynkin's isomorphism

Assume X is locally compact and separable, and that functions of H are continuous. By the Banach-Steinhaus theorem, given any point $x \in X$, there exists an element of H, denoted η_x defined by the identity: $f(x) = \langle \eta_x, f \rangle_H$. Note that $\eta_x = \sum e_k(x) e_k$ for any orthonormal basis of H.

The resolvent V_λ is necessarily Fellerian and induces a strong Markov process. Denote by l_t^x the local time at x of this Markov process. Let x and y be two points of X. Set: $\langle \eta_x, \eta_y \rangle_H = \sum e_k(x) e_k(y) = K(x,y)$ so that $A^{-1} f(x) = \int K(x,y) f(y) m(dy)$ or $A^{-1} f = \int \eta_y f(y) m(dy)$.

Set $V_0(x,y) = \langle (I+B)^{-1} \eta_x, \eta_y \rangle_H$ and note that $V_0(x,y)$ is a kernel for V_0. Indeed, for any $f, g \in \mathbb{L}^2(m)$, $\langle V_0 f, g \rangle_{\mathbb{L}^2(m)} = \langle (I-B)^{-1}(-A)^{-1} f, g \rangle_{\mathbb{L}^2(m)} = \langle (I-B)^{-1}(-A)^{-1} f, A^{-1} g \rangle_H = \int f(x) g(y) V_0(x,y) m(dx) m(dy)$.

As a consequence, $V_0(x,y) = \mathbb{E}_x(l_\zeta^y)$.

Applying the construction of the section 3.1, we see the kernel $K(x,y)$ is the covariance of a Gaussian process $(Z_x = \psi(\eta_x) = \sum e_k(x) w_k, x \in X)$, for any ONB e_k of H.

More generally, for any non-negative finitely supported measure $\chi = \sum_1^N p_j \delta_{u_j}$ on X, letting C be the finite rank operator: $C = \sum_1^N p_j \eta_{u_j} \otimes \eta_{u_j}$, set $V_\chi(x,y) = \langle (I - B + C)^{-1} \eta_x, \eta_y \rangle_H$

In a similar way as above for V_0, we have $V_\chi(x,y) = \mathbb{E}_x(e^{-\int l_t^z \chi(dz)} dl_t^y)$. Then, from section 3-1

$$E(Z_x \overline{Z}_y) e^{-\frac{1}{2}(\langle (C-B)\psi, \overline{\psi}\rangle_H)} = E\left(e^{-\frac{1}{2}(\langle (C-B)\psi, \overline{\psi}\rangle_H)}\right) \langle (I-B+C)^{-1} \eta_x, \eta_y \rangle_H$$

$$= E\left(e^{-\frac{1}{2}(\langle (C-B)\psi, \overline{\psi}\rangle_H)}\right) V_\chi(x,y)$$

But $\langle C\psi, \overline{\psi} \rangle_H = \int \psi(\eta_u) \overline{\psi}(\eta_u) \chi(du) = \int Z_u \overline{Z}_u \chi(du)$ and $\langle B\psi, \overline{\psi} \rangle_H = \left\langle (-A)^{-1} \frac{L-\widehat{L}}{2} \psi, \overline{\psi} \right\rangle_H = \sum \left\langle (-A)^{-1} \frac{L-\widehat{L}}{2} e_k, e_l \right\rangle_H w_k^1 w_l^2$.

On the other hand, at least formally in general but exactly in the finite dimensional case,

$$\int \frac{L - \widehat{L}}{2} Z_u \overline{Z}_u m(du) = \left\langle (-A)^{-1} \frac{L - \widehat{L}}{2} Z, \overline{Z} \right\rangle_H$$

$$= \sum w_k^1 w_l^2 \left\langle (-A)^{-1} \frac{L - \widehat{L}}{2} e_k, e_l \right\rangle_H$$

Hence we can denote: $\langle B\psi, \overline{\psi} \rangle_H$ by $\left\langle \frac{L-\widehat{L}}{2} Z, \overline{Z} \right\rangle_{L^2(m)}$. Therefore

$$E\left(Z_x \overline{Z}_y e^{\frac{1}{2}\left\langle \frac{L-\widehat{L}}{2} Z, \overline{Z} \right\rangle_{L^2(m)} - \int Z_u \overline{Z}_u \chi(du)} \right) =$$

$$E \otimes \mathbb{E}_x \left(\int_0^\zeta e^{\frac{1}{2}\left\langle \frac{L-\widehat{L}}{2} Z, \overline{Z} \right\rangle_{L^2(m)}} e^{-\int l_t^z \chi(dz) - \int Z_u \overline{Z}_u \chi(du)} dl_t^y \right)$$

Let $\mathbb{E}_{x,y}^t$ denote the non-normalized law of the bridge of duration t from x to y. Set $\mu_{x,y} = \int_0^\infty \mathbb{E}_{x,y}^t dt$, so that $\mu_{x,y}(1) = \mathbb{E}_x(l_\zeta^y) = V_0(x,y)$ and denote l^z the local time at z for the duration of the bridge.
Then

$$E\left(Z_x \overline{Z}_y e^{\frac{1}{2}\left\langle \frac{L-\widehat{L}}{2} Z, \overline{Z} \right\rangle_{L^2(m)} - \int Z_u \overline{Z}_u \chi(du)} \right) =$$

$$E\left(e^{\frac{1}{2}\left\langle \frac{L-\widehat{L}}{2} Z, \overline{Z} \right\rangle_{L^2(m)} - \int Z_u \overline{Z}_u \chi(du)} \int e^{-\int l^z \chi(dz)} d\mu_{x,y} \right)$$

Finally, we get that for any continuous bounded function F of N non negative real coordinates, and any N-uple of points u_j in X,

$$E\left(Z_x \overline{Z}_y e^{\frac{1}{2}\left\langle \frac{L-\widehat{L}}{2} Z, \overline{Z} \right\rangle_{L^2(m)}} F(Z_{u_j} \overline{Z}_{u_j}) \right) =$$

$$E \otimes \mu_{x,y} \left(e^{\frac{1}{2}\left\langle \frac{L-\widehat{L}}{2} Z, \overline{Z} \right\rangle_{L^2(m)}} F(l^{u_j} + Z_{u_j} \overline{Z}_{u_j}) \right)$$

and the formula finally extends to

Proposition 3.1. *For any bounded measurable function of a real field on X:*

$$(*bis) \quad E\left(Z_x \overline{Z}_y e^{\frac{1}{2}\left\langle \frac{L-\widehat{L}}{2} Z, \overline{Z} \right\rangle_{L^2(m)}} F(Z\overline{Z}) \right) =$$

$$E \otimes \mu_{x,y} \left(e^{\frac{1}{2}\left\langle \frac{L-\widehat{L}}{2} Z, \overline{Z} \right\rangle_{L^2(m)}} F(l + Z\overline{Z}) \right)$$

It induces the formula (*) obtained in the finite case if we consider the trace of the process on any finite subset.

We see also that the restriction of the twisted Gaussian measure
$$e^{\frac{1}{2}\left\langle \frac{L-\hat{L}}{2} Z, \overline{Z} \right\rangle_{L^2(m)}} P(dZ)$$
to $\sigma(Z_u \overline{Z}_u, u \in X)$ is a probability measure Q under which the distribution of the process $(Z_u \overline{Z}_u, u \in X)$ is infinitely divisible. Moreover, the important fact is that this probability is absolutely continuous with respect to the restriction of P to $\sigma(Z_u \overline{Z}_u, u \in X)$. It is clear that that formula (**) of the corollary extends in the same way to yield the following

Corollary 3.1. *For any bounded measurable function of a nonnegative field on X :*

$$(**bis) \quad \int \rho_x F(\rho_u, u \in X) Q(d\rho) = \int \mu_{x,x}(F(l^u + \rho_u, u \in X)) Q(d\rho)$$

Hence it follows for example that the continuity of the Gaussian field Z implies the continuity of the local time field l under all loop measures μ_{xx}.

Note finally that these results, as in the symmetric case, can be extended to some situations where the local time does not exist (like the two dimensional Brownian motion), by considering the centered occupation field and the "Wick square" : $Z_x \overline{Z}_x$: (formally given by $Z_x \overline{Z}_x - K(x,x)$) as generalized random fields. This makes sense for the Wick square provided K is a Hilbert-Schmidt operator.

References

1. S. Bochner. Completely monotone functions on partially ordered spaces. *Duke Math. J.* **9** (1942), 519–526.
2. E. B. Dynkin. *Local times and Quantum fields* (Seminar on Stochastic processes, Gainesville 1982). Progr. Prob. Statist., Birkhäuser, 7 (1984), 69–84.
3. N. Eisenbaum, H. Kaspi. *On permanental processes.* ArXiv preprint math/0610600.
4. Y. Le Jan. On the Fock space representation of functionals of the occupation field and their renormalization. *Journal of Functional Analysis* **80** (1988), 88–108.
5. Y. Le Jan. *On the Fock space representation of occupation times for non reversible Markov processes* (Stochastic Analysis, Paris 1987). Lecture Notes in Mathematics, Springer, 1322 (1988), 134–138.
6. M. B. Marcus, J. Rosen. Sample path properties of the local times of strongly symmetric Markov processes via Gaussian processes. *Ann. Prob.* **20** (1992), 1603–1684.

7. B. Simon. Trace ideals and their applications. *London Math. Soc. Lect. Notes*, Cambridge, **35** (1979).
8. D. Vere Jones. Alpha permanents and their applications to multivariate gamma, negative binomial and ordinary binomial distributions. *New Zeland J. Math.* **26** (1997), 125–149.

Large deviations for the two-dimensional Yang-Mills measure

Thierry Lévy

Département de Mathématiques et Applications,
École Normale Supérieure, CNRS,
Paris, France
E-mail: levy@dma.ens.fr

This article presents a large deviations principle for the two-dimensional Yang-Mills measure, which is the first mathematical result relating the Yang-Mills measure on a compact surface and the Yang-Mills action. Most of the material is issued from a paper[8] written in collaboration with James Norris. In the present survey, we have tried to give a simple presentation of the problem and its solution, in particular by emphasizing the analogy between the Yang-Mills situation and the more familiar large deviations results for the Brownian motion or Brownian bridges.

Keywords: Large deviations, Cameron-Martin space, connections, random holonomy process

1. The Yang-Mills measure as a Gibbs measure

1.1. *The Brownian motion*

Let E be a finite-dimensional Euclidean vector space. Let $q : [0,1] \longrightarrow E$ be a smooth trajectory starting at 0. From the point of view of the classical dynamics of a particle evolving in the empty space E, the most natural quantity associated to q is its action $S(q)$ defined by

$$S(q) = \int_0^1 \parallel \dot{q}(t) \parallel^2 \, dt.$$

It is a non-negative real number and $S(q) = 0$ if and only if $\dot{q}(t) = 0$ for all $t \in [0,1]$. More interestingly, the square root of S is a pre-Hilbertian norm on the space $C_0^\infty([0,1], E)$ of smooth trajectories starting at 0. The associated Hilbert space is the Sobolev space $W_0^{1,2}([0,1], E)$ of absolutely continuous trajectories starting at 0 whose almost-everywhere defined derivative is square-integrable.

The Brownian motion with values in E starting at 0 can perfectly be defined without any reference to the Sobolev space $W_0^{1,2}([0,1],E)$. Nevertheless, it is well known that these two objects are deeply related to each other, for instance through the Cameron-Martin quasi-invariance principle. Another relation is revealed by Schilder's large deviation principle.[2] Let W denote the Wiener measure on $C_0^0([0,1],E)$. The inclusion $W_0^{1,2} \subset C_0^0$ allows us to extend the action S to a functional on $C_0^0([0,1],E)$, by setting $S(f) = +\infty$ if $f \notin W_0^{1,2}$.

Theorem 1.1 (Schilder's LDP). *Endow $C_0^0([0,1],E)$ with the topology of uniform convergence. Let $A \subset C_0^0([0,1],E)$ be an element of the cylinder σ-field. For each $\lambda \in \mathbb{R}$, set $\lambda A = \{\lambda q : q \in A\}$. Then*

$$-\inf_{\mathring{A}} \frac{1}{2}S \leq \varliminf_{T \to 0} T \log \mathsf{W}\left(\frac{1}{\sqrt{T}}A\right) \leq \varlimsup_{T \to 0} T \log \mathsf{W}\left(\frac{1}{\sqrt{T}}A\right) \leq -\inf_{\overline{A}} \frac{1}{2}S.$$

This theorem suggests that W is a kind of Gibbs measure on $C_0^0([0,1],E)$ associated to the energy $\frac{1}{2}S$:

$$\mathsf{W}(dq) \approx \frac{1}{Z} e^{-\frac{S(q)}{2}} dq,$$

although of course the right-hand side is meaningless.

The Yang-Mills measure YM is a probability measure of the same kind as the Wiener measure. It is thought of by physicists as the Gibbs measure of an action S_{YM} called the Yang-Mills action. On the other hand, it is constructed at a mathematical level of rigour without any reference to this action. The large deviation principle which we present in this paper relates the measure YM and the action S_{YM}.

1.2. The Yang-Mills action

Let M be an oriented compact two-dimensional manifold, possibly with boundary, which plays the role of space-time. Let us endow M with a Borel measure vol which is equivalent to Lebesgue measure in any chart. This measure can be identified with a smooth 2-form on M. For example, let us take for M the torus $\mathbb{T}^2 = \mathbb{R}^2/\mathbb{Z}^2$ on which we have the coordinates (x,y), endowed with the Lebesgue measure vol $= dxdy$ which we identify with the 2-form $dx \wedge dy$.

Let G be a compact Lie group whose Lie algebra $\mathfrak{g}$ is endowed with an invariant scalar product. For example, let us take $G = SU(2)$. The Lie algebra of $SU(2)$ is the real vector space

$$\mathfrak{su}(2) = \left\{ X_{\alpha,\beta,\gamma} = \begin{pmatrix} i\alpha & \gamma + i\beta \\ -\gamma + i\beta & -i\alpha \end{pmatrix} : \alpha, \beta, \gamma \in \mathbb{R} \right\}.$$

It is endowed with the norm $\| X_{\alpha,\beta,\gamma} \|^2 = -\text{Tr}(X_{\alpha,\beta,\gamma}^2) = \alpha^2 + \beta^2 + \gamma^2$, which is invariant under the adjoint action of $SU(2)$.

The set on which the Yang-Mills action is defined is the set of smooth connections on a principal G-bundle over M. Locally, a connection is a 1-form on M with values in the Lie algebra of G and it turns out that most of the issues we are dealing with are local. We postpone a discussion of global issues to the last section. Keeping in mind that this is a simplification, let us define the set $\mathcal{A}$ of smooth connections as the set of smooth $\mathfrak{g}$-valued 1-forms on M. In our example,

$$\mathcal{A} = \left\{ \omega = \omega_1 \, dx + \omega_2 \, dy : \omega_1, \omega_2 \in C^\infty(\mathbb{T}^2, \mathfrak{su}(2)) \right\}.$$

Example 1.1. Let $u : \mathbb{T}^2 \longrightarrow SU(2)$ be a smooth function. Then u determines a connection, denoted by $u^{-1}du$ and defined by

$$u^{-1}du = u^{-1}\frac{\partial u}{\partial x} \, dx + u^{-1}\frac{\partial u}{\partial y} \, dy.$$

A connection of the form $u^{-1}du$ is called a flat connection. Now how do you tell that a connection is flat? Recall that a classical real 1-form $\alpha = \alpha_1 dx + \alpha_2 dy$ is called exact if it is of the form df for some smooth function f. Flat connections are non-linear analogues of exact 1-forms. Poincaré's lemma states that, locally, the 1-form α above is exact if and only if $\frac{\partial \alpha_2}{\partial x} - \frac{\partial \alpha_1}{\partial y} = 0$. The quantity which plays the rôle of $\frac{\partial \alpha_2}{\partial x} - \frac{\partial \alpha_1}{\partial y}$ for connections is called the curvature.

Let $\omega = \omega_1 \, dx + \omega_2 \, dy$ be a connection on $\mathbb{T}^2$. The curvature of ω is the function on $\mathbb{T}^2$ with values in $\mathfrak{su}(2)$ defined by

$$\Omega = \frac{\partial \omega_2}{\partial x} - \frac{\partial \omega_1}{\partial y} + [\omega_1, \omega_2].$$

The bracket denotes the Lie bracket, which in our description of $\mathfrak{su}(2)$ is given simply by the commutator $[X, Y] = XY - YX$. It is a theorem that, at least locally, ω is of the form $u^{-1}du$ if and only if $\Omega = 0$.

Definition 1.1 (Yang-Mills action). *Let ω be an element of $\mathcal{A}$. The Yang-Mills action of ω is defined by*

$$S_{\text{YM}}(\omega) = \int_{\mathbb{T}^2} \| \Omega \|^2 \, dxdy.$$

The connections whose action is equal to zero are the flat connections. At this point, we must mention the action of the gauge group on the set $\mathcal{A}$. It is an infinite-dimensional group which acts by affine transformations similar to the transformations $\alpha \rightsquigarrow \alpha + df$ on real 1-forms. This action preserves the Yang-Mills action and the fundamental object should be an orbit of the gauge group rather than a connection. In particular, any two flat connections are locally gauge-equivalent. Nevertheless, the space of orbits of flat connections is far from trivial. It is a complicated finite-dimensional orbifold whose geometric structure is not yet fully understood.[1,11] Fortunately, this structure can be ignored in the study of the large deviation principle.

The presence of a quadratic term in the definition of Ω makes the search for a space of connections with finite energy less straightforward than in the Brownian case. Unless there is an unexpected cancellation between different terms, a connection needs to have its components in L^4 and the derivatives of its components in L^2 in order to have a finite action. It turns out that, since M is 2-dimensional, the inclusion $W^{1,2}(M, \mathfrak{g}) \subset L^p(M, \mathfrak{g})$ holds for all $p < \infty$, in particular for $p = 4$. It is interesting to note that the inclusion would hold for $p \leq 6$ if M was 3-dimensional and $p \leq 4$ if M was 4-dimensional.

The forthcoming results will confirm that the space $W^{1,2}\mathcal{A}$ of connections whose components are in the Sobolev space $W^{1,2}$ plays the rôle of the Cameron-Martin space for the Yang-Mills action.

Remark 1.1. There are two natural non-negative functionals on $W^{1,2}\mathcal{A}$: the Yang-Mills action S_{YM} and the $W^{1,2}$ norm. Since the injection $W^{1,2} \hookrightarrow L^4$ is continuous, there exists a constant K, which depends on the definition one chooses for the $W^{1,2}$ norm, such that

$$\forall \omega \in W^{1,2}\mathcal{A},\ S_{\mathsf{YM}}(\omega) \leq K \parallel \omega \parallel_{1,2}^2 .$$

On the other hand, the gauge group acts on $W^{1,2}\mathcal{A}$ and preserves the Yang-Mills action. This action does not at all preserve the $W^{1,2}$ norm, it can in fact increase it arbitrarily. Hence,

$$\not\exists L > 0, \forall \omega \in W^{1,2}\mathcal{A},\ \parallel \omega \parallel_{1,2}^2 \leq L\, S_{\mathsf{YM}}(\omega).$$

This result of non-existence is extremely annoying. The natural measure of the energy on $W^{1,2}\mathcal{A}$ is the Yang-Mills action. On the other hand, the $W^{1,2}$ norm enjoys fundamental topological properties like the weak precompactness of bounded sets. It would be very desirable to have a precompactness result for sets of connections with bounded action. Such a result has been proved by Karen Uhlenbeck[12] in 1982.

Theorem 1.2 (Uhlenbeck compactness). *Let $(\omega_n)_{n\geq 0}$ be a sequence of elements of $W^{1,2}\mathcal{A}$ such that the sequence $(S_{\mathsf{YM}}(\omega_n))_{n\geq 0}$ is bounded. Then there exists $\omega \in W^{1,2}\mathcal{A}$ such that, after extracting some subsequence of $(\omega_n)_{n\geq 0}$ and letting a (possibly different) gauge transformation act on each of its terms, one has*

$$\omega_n \xrightarrow{W^{1,2}} \omega.$$

The proof of this beautiful and difficult theorem goes in two main steps. First, one proves that a connection with small Yang-Mills action is, *locally*, gauge-equivalent to a connection with small $W^{1,2}$ norm. Second, one applies this result to each term of the sequence of connections, by patching local domains on which it holds. On the overlaps of the domains, one has to deal with distinct gauge transformations but it is possible to control the difference between them and, by compactness arguments, to guarantee the existence of a subsequence along which they behave nicely. In addition to the original article of K. Uhlenbeck,[12] there is a self-contained and detailed proof of this theorem and other related ones in a monograph by K. Wehrheim.[14]

1.3. *The necessity to consider holonomy*

Let us think as if the curvature was a linear functional of the connection, as it is in fact when G is Abelian, for example $G = U(1)$. In this case, the Gibbs measure of the Yang-Mills action should be the Gaussian measure on the Hilbert space $W^{1,2}\mathcal{A}$. The classical theory of Gaussian measures asserts that this measure is supported by $W^{k,2}\mathcal{A}$ if and only if k is such that the inclusion $W^{1,2} \hookrightarrow W^{k,2}$ is Hilbert-Schmidt. This happens exactly when $k < 0$. In other words, this Gaussian measure is not supported by a space of functions, but by a space of distributions. Thus, it is not to be expected that a typical connection under the Gibbs measure of the Yang-Mills action be even an L^2 connection. In fact, pointwise evaluation is not even defined on $W^{1,2}$, since $W^{1,2} \not\subset C^0$ in two dimensions.

The way around this previsible difficulty is given by a natural set of observables on $W^{1,2}\mathcal{A}$ which is both natural geometrically and smoother in that it requires less regularity on the connection to be defined. It is the holonomy along smooth curves.

Let $c : [0,1] \longrightarrow M$ be a smooth curve. Let ω be a $W^{1,2}$ connection on M. Although the components of ω have no pointwise meaning, they have a trace along c, which is a $W^{\frac{1}{2},2}$ function, in particular an L^2 function. In other words, the function

$$t \mapsto \omega(\dot{c}(t)) : [0,1] \longrightarrow \mathfrak{g}$$

is well defined and belongs to $L^2([0,1],\mathfrak{g})$. It is thus possible to solve the differential equation

$$\begin{cases} h_t^{-1}\dot{h}_t = -\omega(\dot{c}(t)) \\ h_0 = 1 \end{cases}$$

whose solution is a function $h : [0,1] \longrightarrow G$. The element h_1 of G is called the holonomy of ω along c and it is denoted by $\mathrm{hol}(\omega,c)$. The holonomy enjoys the following natural multiplicativity property: if c_1 and c_2 can be concatenated into c_1c_2, then $\mathrm{hol}(\omega, c_1c_2) = \mathrm{hol}(\omega, c_1)\mathrm{hol}(\omega, c_2)$.

Note that if G is Abelian, then the holonomy along c can be expressed simply as

$$\mathrm{hol}(\omega, c) = \exp - \int_c \omega.$$

The set of observables $\mathrm{hol}(\cdot, c)$, where c spans the set of smooth curves on M is complete. This means that a connection ω is characterised, up to the action of the gauge group to which we have made a short reference earlier, by the elements $\mathrm{hol}(\omega, c)$ of G.

Let us think by analogy with the Brownian case again. If f is a square-integrable real function on $[0,1]$, then $\int_0^1 f(t)\,\dot{q}(t)$ is well defined for every $q \in W^{1,2}$. Then, despite the fact that this integral is not defined in general when $q \in C^0$, the Wiener integral $\int_0^1 f(t)\,dB_t$ is defined outside a negligible set.

Transposing this to the Yang-Mills situation, we find that, although $\mathrm{hol}(\omega, c)$ is not defined in general for $\omega \in W^{k,2}$ with $k < 0$, we could expect it to be defined in some sense, almost surely under the Gibbs measure of the Yang-Mills energy.

Let us summarise this section by a dictionary between the Brownian and Yang-Mills situations.

Brown		Yang-Mills
$q : [0,1] \longrightarrow \mathbb{R},$ $q(0) = q(1) = 0$	random object	$\omega = \omega_1\, dx + \omega_2\, dy,$ $\omega_1, \omega_2 : \mathbb{T}^2 \longrightarrow \mathfrak{su}(2)$
$\dot{q}(t)$	derivative	$\Omega = \partial_x \omega_2 - \partial_y \omega_1 + [\omega_1, \omega_2]$
$\int_0^1 |\dot{q}(t)|^2\, dt$	energy	$\int_{\mathbb{T}^2} \| \Omega(x) \|^2\, \text{vol}(dx)$
$\forall t \in [0,1], q(t) = 0$	zero energy	$\omega = u^{-1} du,$ $u : \mathbb{T}^2 \longrightarrow SU(2)$ (flat)
$q \mapsto q(t)$ for $t \in [0,1]$	observable	$\omega \mapsto \text{hol}(\omega, c)$ for $c \in \mathsf{P}(M)$
$(B_t)_{t \in [0,1]}$	stochastic process	$(H_c)_{c \in \mathsf{P}(M)}$

2. The Yang-Mills measure as a stochastic process

Now that we have discussed the heuristics of the Yang-Mills measure, we are going to describe a way to construct it. The mathematical work on the two-dimensional Yang-Mills measure as a probabilistic object has been initiated by L. Gross[4] in the late 1980's and strongly stimulated by a seminal paper of E. Witten.[15] B. Driver[3] then clarified the case where the base space is a Euclidean plane, and A. Sengupta[10] gave the first construction of the measure on an arbitrary bundle over an arbitrary compact surface, by an infinite-dimensional approach. The construction presented below is due to the author[6,7] and is based on a procedure of finite-dimensional approximation.

2.1. Finite-dimensional distributions

Let us start again with the data of the compact surface M and the measure vol. In order to make things simpler, let us endow M with a Riemannian metric. For example, in the case of the torus $\mathbb{T}^2$, we choose the metric $dx^2 + dy^2$, whose Riemannian volume is precisely $\text{vol} = dxdy$.

Let $\mathsf{P}(M)$ denote the set of rectifiable curves $c : [0,1] \longrightarrow M$ parametrized at constant speed. The Yang-Mills measure is the distribution of a collection of G-valued random variables $(H_c)_{c \in \mathsf{P}(M)}$ indexed by $\mathsf{P}(M)$. Moreover, this collection is multiplicative in the sense that $H_{c_1 c_2} = H_{c_2} H_{c_1}$ almost surely whenever c_1 and c_2 can be concatenated. The previous section explains why such an object could play the rôle of the hypothetical Gibbs measure of the Yang-Mills action, but we will not refer to this discussion during the construction. Instead, we would like to convey the following

Large deviations for the two-dimensional Yang-Mills measure 61

intuitive idea : the Yang-Mills measure is the distribution of a Brownian bridge on G indexed by the set of curves on M rather than an interval of $\mathbb{R}$.

In order to construct the Yang-Mills process, we need to specify its finite-dimensional distributions. In the case of the Brownian bridge indexed by $[0,1]$, the first step would be to choose a finite subset of $[0,1]$. This finite subset splits $[0,1]$ into a finite number of disjoint sub-intervals. The analogue operation on M is to split M into a finite number of smaller pieces, by drawing a graph on it. Loosely speaking, a graph is a set of injective curves, called edges, which satisfy a few simple conditions on their intersections. These edges determine faces, and we assume that these faces are topologically trivial : we assume that they are homeomorphic to disks. For instance, a single edge in the middle of $\mathbb{T}^2$ does not constitute an acceptable graph.

Let $\mathbb{G} = (\mathbb{V}, \mathbb{E}, \mathbb{F})$ be a graph. This notation means that $\mathbb{V}, \mathbb{E}, \mathbb{F}$ are respectively the sets of vertices, edges and faces. The distribution of $(H_e)_{e \in \mathbb{E}}$ is a probability measure on $G^{\mathbb{E}}$. In order to guess what this distribution is, let us pursue the analogy with the Brownian bridge on $[0,1]$. In the Brownian case, the finite-dimensional distribution corresponding to $T = \{t_1 < \ldots < t_n\}$ has a density with respect to the Lebesgue measure on $\mathbb{R}^n$, and this density is a product of factors, one for each interval. For each interval $[t_k, t_{k+1}]$, the factor evaluated at a vector $(q_1, \ldots, q_n) \in \mathbb{R}^n$ is the heat kernel at a time $t_{k+1} - t_k$ which is the length of the interval, evaluated at the difference $q_{k+1} - q_k$ of the values of the vector at the end points of the interval.

In the Yang-Mills situation, $\mathbb{E}$ plays the rôle of T and each face $F \in \mathbb{F}$ that of an interval. The Lebesgue measure is replaced by the Haar measure on $G^{\mathbb{E}}$. We expect the distribution of $(H_e)_{e \in \mathbb{E}}$ to have a density with respect to the Haar measure and this density should be a product of factors, one for each face. Let F be a face and let $g = (g_e)_{e \in \mathbb{E}}$ be an element of $G^{\mathbb{E}}$. The factor associated to F evaluated at g should be a heat kernel, at a time which is now the area $\text{vol}(F)$ of F. It remains to decide what is the analogue of $q_{k+1} - q_k$, that is, at which point of G to evaluate the heat kernel. The formal linear combination of points $\{t_{k+1}\} - \{t_k\}$ is the oriented boundary of the interval $[t_k, t_{k+1}]$. In the same vein, the boundary of F is a cycle in the graph $\mathbb{G}$. Provided M is oriented, this cycle is also oriented. It still lacks an origin. Let us choose one and write this path $e_{i_1}^{\epsilon_1} \ldots e_{i_k}^{\epsilon_k}$. Here, $\epsilon_i = \pm 1$, to take care of the orientation of the edges. What should replace $q_{k+1} - q_k$ is now $g_{e_{i_k}}^{\epsilon_k} \ldots g_{e_{i_1}}^{\epsilon_1}$, which we denote by $h_{\partial F}(g)$. This element of

G is defined only up to conjugation because we have chosen an origin for the cycle ∂F. Fortunately, the heat kernel on G is invariant by conjugation. Finally, denoting by $(Q_t)_{t>0}$ the heat kernel on G, we find the following expression for the distribution of $(H_e)_{e \in \mathbb{E}}$:

$$\mathsf{YM}_T^{\mathbb{G}}(dg) = \frac{1}{Z_T^{\mathbb{G}}} \prod_{F \in \mathbb{F}} Q_{T\mathrm{vol}(F)}(h_{\partial F}(g)) \, dg.$$

A scaling factor $T > 0$ has appeared in this expression: we have replaced the measure vol by Tvol. The large deviation principle will hold in the limit $T \to 0$. The constant $Z_T^{\mathbb{G}}$ is the normalisation constant and it ensures that the measure $\mathsf{YM}_T^{\mathbb{G}}$ is a probability measure.

Let us summarise again the analogy between the Brownian case and the Yang-Mills case by a dictionary.

Brown		Yang-Mills
$[0,1]$	index set	$\mathsf{P}(M)$
$T = \{t_1 < \ldots < t_n\} \subset [0,1]$	subdivision	$\mathbb{F}$: faces of $\mathbb{G} = (\mathbb{V}, \mathbb{E}, \mathbb{F})$
$\mathbb{R}^T \ni q = (q_1, \ldots, q_n)$	configuration space	$G^{\mathbb{E}} \ni g = (g_e)_{e \in E}$
$dq = \bigotimes_{i=1}^n dq_i$	uniform measure	$dg = \bigotimes_{e \in E} dg_e$
$\prod_{i=0}^n D_{[t_i, t_{i+1}]}(q)$	density	$\prod_{F \in \mathbb{F}} D_F(g)$
$\frac{1}{\sqrt{2\pi t}} e^{-\frac{x^2}{2t}}$	heat kernel	$Q_t(x)$
$\partial([t_i, t_{i+1}]) = \{t_{i+1}\} - \{t_i\}$	boundary	$\partial F = e_{i_1}^{\epsilon_1} \ldots e_{i_n}^{\epsilon_n}$
$q_{t_{i+1}} - q_{t_i}$		$h_{\partial F}(g) = g_{e_{i_n}}^{\epsilon_n} \ldots g_{e_{i_1}}^{\epsilon_1}$
$|t_{i+1} - t_i|$	length / area	$\mathrm{vol}(F)$
$\frac{1}{\sqrt{2\pi T|t_{i+1}-t_i|}} e^{-\frac{(q_{i+1}-q_i)^2}{2T|t_{i+1}-t_i|}}$	D	$Q_{T\mathrm{vol}(F)}(h_{\partial F}(g))$
$\frac{1}{\sqrt{2\pi T}}$	normalisation	$Z_T^{\mathbb{G}}$

2.2. The Yang-Mills process

The construction of the process from the finite-dimensional distributions presented in the last section poses a problem which is specific to the Yang-Mills situation. In fact, the measures $\mathsf{YM}_T^{\mathbb{G}}$ as $\mathbb{G}$ spans the set of graphs on M do not suffice to describe all finite-dimensional distributions of a process indexed by $\mathsf{P}(M)$. Indeed, even a single rectifiable path on M cannot in general be written as a finite concatenation of edges. Therefore, it is necessary to perform a kind of approximation. The topology that we use on P(M) is the following. We say that a sequence $(c_n)_{n\geq 0}$ of paths converges to c if the length of c_n converges to that of c and the sequence converges uniformly when all the paths are parametrized at constant speed. We also insist that all c_n share the same endpoints as c.

The canonical space of the process is the set of multiplicative functions from P(M) to G: we say that h : P(M) $\longrightarrow$ G is multiplicative if $h(c^{-1}) = h(c)^{-1}$ for all $c \in$ P(M) and $h(c_1 c_2) = h(c_2) h(c_1)$ whenever $c_1 c_2$ exists. We denote by $\mathcal{M}$(P(M), G) the set of multiplicative functions from P(M) to G.

Theorem 2.1 (L.[6]). *For each $T > 0$, there exists a unique probability measure* YM_T *on the cylinder σ-field of* $\mathcal{M}$(P(M), G) *such that, under* YM_T, *the canonical process* $(H_c)_{c \in \mathsf{P}(M)}$ *satisfies the following properties.*

1. *For all graph* $\mathbb{G} = (\mathbb{V}, \mathbb{E}, \mathbb{F})$, *the distribution of* $(H_e)_{e \in \mathbb{E}}$ *is* $\mathsf{YM}_T^\mathbb{G}$.
2. *When a sequence* $(c_n)_{n \geq 0}$ *of paths converges towards c with fixed endpoints,* H_{c_n} *converges in probability to H_c.*

Remark 2.1. With the definition of P(M) we have adopted here, this theorem has not yet been published. The articles to which we refer deal only with piecewise smooth curves, which is enough for most purposes, including those of the work presented here. The case of rectifiable curves will be treated in a forthcoming paper.

3. The large deviations principle

The discussion in the first section of this paper explains how the holonomy determines a mapping $W^{1,2}\mathcal{A} \longrightarrow \mathcal{M}(\mathsf{P}(M), G)$ which sends ω to $\mathrm{hol}(\omega, \cdot)$. We will ignore the fact that the paths of P(M) are rougher than those which we considered in our description of the holonomy. The main point is that, up to the action of the gauge group, which we also choose to ignore as far as possible, this mapping is one-to-one. This allows us to define a functional on $\mathcal{M}(\mathsf{P}(M), G)$ as follows:

$$\forall f \in \mathcal{M}(\mathsf{P}(M), G),\ I_{\mathsf{YM}}(f) = \begin{cases} \frac{1}{2} S_{\mathsf{YM}}(\omega) & \text{if } f(\cdot) = \mathrm{hol}(\omega, \cdot) \text{ for } \omega \in W^{1,2}\mathcal{A} \\ +\infty & \text{otherwise.} \end{cases}$$

Theorem 3.1 (L., Norris[8]). *As T tends to 0, the family of probability measures* $(\mathsf{YM}_T)_{T>0}$ *satisfies a large deviation principle of speed T and rate I_{YM}.*

The topology on $\mathcal{M}(\mathsf{P}(M), G)$ is the trace of the product topology on $G^{\mathsf{P}(M)}$, which is compact. In particular, I_{YM} is a good rate function, but beware that $\mathcal{M}(\mathsf{P}(M), G)$ is not a Polish space. The signification of the theorem is the following: for every $A \subset \mathcal{M}(\mathsf{P}(M), G)$ element of the cylinder σ-field,

$$-\inf_{\overset{\circ}{A}} I_{\mathsf{YM}} \leq \varliminf_{T \to 0} T \log \mathsf{YM}_T(A) \leq \varlimsup_{T \to 0} T \log \mathsf{YM}_T(A) \leq -\inf_{\overline{A}} I_{\mathsf{YM}}.$$

In the rest of this section, we give an overview of the proof. It goes in three main steps:

- large deviations of the finite-dimensional distributions,
- projective limit of the finite-dimensional large deviation principles,
- identification of the rate function.

1. The finite-dimensional problem relies fully on the classical large deviations result for the heat kernel on G, or more generally on a compact Riemannian manifold.[13]

Theorem 3.2. *Let $Q_t(\cdot, \cdot)$ be the heat kernel on a compact Riemannian manifold G. Then, uniformly for all $x, y \in G$, one has*

$$\lim_{t \to 0} -2t \log Q_t(x, y) = d(x, y)^2.$$

In our definition of the Yang-Mills measure, we have used the notation Q_t with only one argument. The relation with the present kernel is given by $Q_t(x) = Q_t(1, x)$, where 1 is the unit element of G.

Using this theorem, it is easy to prove that, given a graph $\mathbb{G}$, the measures $\mathsf{YM}_T^{\mathbb{G}}$ satisfy on $G^{\mathbb{E}}$ a large deviation principle as $T \to 0$, with speed T and rate

$$\forall g \in G^{\mathbb{E}}, \ I_{\mathsf{YM}}^{\mathbb{G}}(g) = \sum_{F \in \mathbb{F}} \frac{d(1, h_{\partial F}(g))^2}{2\mathrm{vol}(F)}.$$

2. The second step consists in applying Dawson-Gärtner theorem and an argument of exponential approximation[2] to obtain a large deviation principle for the measures YM_T themselves. One finds something of the following form: the measures YM_T satisfy on $\mathcal{M}(\mathsf{P}(M), G)$ a large deviation principle as $T \to 0$, with speed T and rate

$$\forall f \in \mathcal{M}(\mathsf{P}(M), G), \ \hat{I}_{\mathsf{YM}}(f) = \sup_{\mathbb{G} \text{ graph on } M} I_{\mathsf{YM}}^{\mathbb{G}}(f_{|\mathbb{E}}).$$

3. The last step consists in proving that the rate function $\hat{I}_{\mathsf{YM}}$ coincides with the function I_{YM}. Let us take a moment to look at the similar problem

which occurs in the proof of Schilder's theorem. At some point, one finds an expression of the rate function of the form:

$$\forall q \in C_0^0([0,1]), \ \hat{I}(q) = \sup_{0 < t_1 < \ldots < t_n < 1} \sum_{i=0}^{n-1} \frac{(q(t_{i+1}) - q(t_i))^2}{2|t_{i+1} - t_i|}.$$

One has then to prove that this is exactly the half of the square $W^{1,2}$-norm of q if $q \in W^{1,2}$ and $+\infty$ otherwise.

If $q \in W^{1,2}$, then by Jensen's formula, $2\hat{I}(q) \leq \| q \|_{1,2}^2$. Then, by using for instance dyadic subdivisions and the martingale convergence theorem, one proves the equality of both sides. After that, it remains to prove that $\hat{I}(q) < +\infty$ implies $q \in W^{1,2}$. Here is one way to do it, which is the analogue of the technique we use in the proof of the theorem.

Assume that $\hat{I}(q) < +\infty$. For each subdivision $\mathcal{D}$ of $[0,1]$, let $q^\mathcal{D}$ denote the piecewise affine interpolation of q at the points of $\mathcal{D}$. The function $q^\mathcal{D}$ has three crucial properties: it agrees with q on the points of $\mathcal{D}$, it belongs to $W^{1,2}$ and its square $W^{1,2}$ norm is exactly $\sum_\mathcal{D} \frac{(q(t_{i+1}) - q(t_i))^2}{|t_{i+1} - t_i|}$. Since $\hat{I}(q) < +\infty$, the family of functions $q^\mathcal{D}$, where $\mathcal{D}$ spans the set of subdivisions of $[0,1]$, is a bounded set of $W^{1,2}$. It is thus possible to find a sequence $(\mathcal{D}_n)_{n \geq 0}$ of subdivisions whose mesh tends to 0 and such that the sequence $(q^{\mathcal{D}_n})_{n \geq 0}$ converges weakly in $W^{1,2}$ towards some $r \in W^{1,2}$. In particular, since the inclusion $W^{1,2} \subset C^0$ is compact, $q^{\mathcal{D}_n}$ converges uniformly towards r. On the other hand, since the mesh of $\mathcal{D}_n$ tends to 0, $q^{\mathcal{D}_n}$ converges uniformly to q. Finally, $q = r$ belongs to $W^{1,2}$.

The two major steps of this argument are the construction of $q^\mathcal{D}$ (as obvious as it seems) and the compactness argument. In the Yang-Mills situation, the compactness property is granted by the difficult theorem of Uhlenbeck (Theorem 1.2). The analogue of the construction of $q^\mathcal{D}$ consists in solving the following problem: given $f \in \mathcal{M}(\mathsf{P}(M), G)$ and a graph $\mathbb{G}$, find a connection in $W^{1,2}\mathcal{A}$ which induces the same holonomy as f on the edges of $\mathbb{G}$ and has exactly the Yang-Mills action $\sum_{F \in \mathbb{F}} \frac{d(1, f(\partial F))^2}{\text{vol}(F)}$. This turns out to be one of the longest steps of the proof. As one could guess by looking at the Brownian case, problems arise in the neighbourhood of the graph, in particular near the vertices, where the connection cannot be differentiable. We start by producing explicitly a connection inside each face of the graph, then in the neighbourhood of the interior of each edge and finally near each vertex. Then, we prove that these locally defined connections can be patched together - technically, we construct the principal bundle at the same time that we construct the connection on it.

4. Disintegration of YM_T according to the topology

The large deviations principle that we have presented so far is really true in this form only if G is simply connected, for example if $G = SU(N)$. In this section, we explain how the result must be formulated when G is not simply connected, for example when $G = U(N)$.

4.1. The fundamental group of $U(N)$

Consider a loop $\gamma : [0,1] \longrightarrow U(N)$ such that $\gamma(0) = \gamma(1) = I_N$. The determinant of γ is a loop $\det(\gamma) : [0,1] \longrightarrow U(1) \subset \mathbb{C}$ based at 1. The index around 0 of $\det(\gamma)$ is an element of $\mathbb{Z}$ which depends only on the homotopy class of γ. In fact, the mapping

$$\pi_1(U(N), I_N) \longrightarrow \mathbb{Z}$$
$$[\gamma] \longmapsto \mathrm{Ind}_0\left(\det(\gamma)\right)$$

is an isomorphism.

Hence, the fundamental group of $U(N)$ is isomorphic to $\mathbb{Z}$ and the class of a loop γ can be identified with the topological index of $\det(\gamma)$ around 0 in the complex plane.

4.2. A homotopic splitting of the Yang-Mills process

Consider a Brownian bridge $(Z_t)_{t \in [0,1]}$ with values in $U(1) \subset \mathbb{C}$. It is natural to decompose the law of Z according to the index around 0 of its trajectory. There is a similar issue with the Yang-Mills process. For simplicity, let us consider $M = S^2$. Recall that $G = U(N)$.

Consider a family of loops $(l_t)_{t \in [0,1]}$ based at the north pole which evolve around the sphere as shown by the picture below. It is understood that l_0 and l_1 are both constant loops at the north pole. As a loop in the space of loops on S^2, the family $(l_t)_{t \in [0,1]}$ is assumed to induce the generator of $\pi_2(S^2) \simeq \mathbb{Z}$ specified by the orientation of S^2.

The $U(N)$-valued process $(H_{l_t})_{t \in [0,1]}$ is indexed by $[0,1]$ and it has the distribution of a Brownian bridge on $U(N)$. For each $k \in \mathbb{Z}$, we will call $\mathsf{YM}_{T,k}$ the distribution of the Yang-Mills process conditioned by the fact that the index of the loop $t \mapsto \det H_{l_t}$ is equal to k. It turns out that this definition does not depend on the particular family $(l_t)_{t \in [0,1]}$ one chooses, given the homotopic constraint it must satisfy.

The disintegration $\mathsf{YM}_T = \sum_{k \in \mathbb{Z}} \mathsf{YM}_{T,k}$ is not just an abstract decomposition: it is possible to write down the finite-dimensional distributions of $\mathsf{YM}_{T,k}$ corresponding to the set of edges of a graph.[7] It can also be proved that $\mathsf{YM}_{T,k}$ and $\mathsf{YM}_{T',k'}$ are mutually singular unless $(T,k) = (T',k')$.[7]

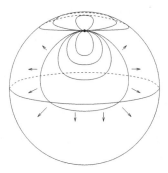

4.3. Topology of principal bundles

As we have mentioned in the first section of this paper, connections are not really differential forms on M. They are in fact differential forms on principal bundles over M. A principal G-bundle over M is a smooth mapping $\pi : P \longrightarrow M$ from a manifold P on which G acts freely on the right. This mapping is required to have locally, that is over open subsets of M which are small enough, the same structure as the projection $U \times G \longrightarrow U$ on the first coordinate.[5] Two principal bundles $\pi : P \longrightarrow M$ and $\pi' : P' \longrightarrow M$ are isomorphic if there is a diffeomorphism $\phi : P \longrightarrow P'$ which intertwines the actions of G and satisfies $\pi' \circ \phi = \pi$. When M is a compact connected oriented surface without boundary and when G is connected, the set of isomorphism classes of principal G-bundles is canonically isomorphic with the fundamental group of G.[9]

For example, there is exactly one isomorphism class of $U(N)$-bundles over S^2 for each integer $k \in \mathbb{Z}$. Thus, up to isomorphism, there is one set of connections over M for each $k \in \mathbb{Z}$, which we denote by $W^{1,2}\mathcal{A}_k$. Now, ignoring again the action of the gauge group, there is for each $k \in \mathbb{Z}$ an injection $W^{1,2}\mathcal{A}_k \hookrightarrow \mathcal{M}(\mathsf{P}(M), U(N))$ and this allows us to define a rate function $I_{\mathsf{YM},k}$ as follows:

$$I_{\mathsf{YM},k}(f) = \begin{cases} \frac{1}{2} S_{\mathsf{YM}}(\omega) & \text{if } f(\cdot) = \text{hol}(\omega, \cdot) \text{ for } \omega \in W^{1,2}\mathcal{A}_k \\ +\infty & \text{otherwise.} \end{cases}$$

Note that the image of the inclusion $W^{1,2}\mathcal{A} \hookrightarrow \mathcal{M}(\mathsf{P}(M), U(N))$ is the disjoint union of the images of the inclusions $W^{1,2}\mathcal{A}_k \hookrightarrow \mathcal{M}(\mathsf{P}(M), U(N))$ for $k \in \mathbb{Z}$.

The refined version of the large deviation principle is now the following.

Theorem 4.1 (L., Norris[8]). *Assume that $G = U(N)$. Then for each $k \in \mathbb{Z}$, as T tends to 0, the family of probability measures $(\mathsf{YM}_{T,k})_{T>0}$ satisfies* a large deviation principle of speed T and rate $I_{\text{YM},k}$.

The theorem is true for any connected compact Lie group G instead of $U(N)$, but then $\mathbb{Z}$ has to be replaced by $\pi_1(G)$.

References

1. Michael F. Atiyah and Raoul Bott. The Yang-Mills equations over Riemann surfaces. *Philos. Trans. Roy. Soc. London Ser. A* **308(1505)** (1983), 523–615.
2. Amir Dembo and Ofer Zeitouni. *Large deviations techniques and applications.* Jones and Bartlett Publishers, Boston, MA, 1993.
3. Bruce K. Driver. YM_2: continuum expectations, lattice convergence, and lassos. *Comm. Math. Phys.* **123(4)** (1989), 575–616.
4. Leonard Gross, Christopher King, and Ambar Sengupta. Two-dimensional Yang-Mills theory via stochastic differential equations. *Ann. Physics* **194(1)** (1989), 65–112.
5. Shoshichi Kobayashi and Katsumi Nomizu. *Foundations of differential geometry.* Vol. I. Wiley Classics Library. John Wiley & Sons Inc., New York, 1996. Reprint of the 1963 original, A Wiley-Interscience Publication.
6. Thierry Lévy. Yang-Mills measure on compact surfaces. *Mem. Amer. Math. Soc.* **166(790)** (2003), xiv+122.
7. Thierry Lévy. Discrete and continuous Yang-Mills measure for non-trivial bundles over compact surfaces. *Probab. Theory Related Fields* **136(2)** (2006), 171–202.
8. Thierry Lévy and James R. Norris. Large deviations for the Yang-Mills measure on a compact surface. *Comm. Math. Phys.* **261(2)** (2006), 405–450.
9. Shigeyuki Morita. *Geometry of differential forms*, volume 201 of Translations of Mathematical Monographs. American Mathematical Society, Providence, RI, 2001.
10. Ambar N. Sengupta. Gauge theory on compact surfaces. *Mem. Amer. Math. Soc.* **126(600)** (1997), viii+85.
11. Ambar N. Sengupta. The volume measure for flat connections as limit of the Yang-Mills measure. *J. Geom. Phys.* **47(4)** (2003), 398–426.
12. Karen K. Uhlenbeck. Connections with L^p bounds on curvature. *Comm. Math. Phys.* **83(1)** (1982), 31–42.
13. Srinivasa R. S. Varadhan. Diffusion processes in a small time interval. Comm. *Pure Appl. Math.* **20** (1967), 659–685.
14. Katrin Wehrheim. *Uhlenbeck compactness.* EMS Series of Lectures in Mathematics. European Mathematical Society (EMS), Zürich, 2004.
15. Edward Witten. On quantum gauge theories in two dimensions. *Comm. Math. Phys.* **141(1)** (1991), 153–209.

69

Laplace operator in networks of thin fibers: Spectrum near the threshold

S. Molchanov and B. Vainberg[*]

Dept. of Mathematics, University of North Carolina at Charlotte,
Charlotte, NC 28223, USA
E-mail: smolchan@uncc.edu | brvainbe@uncc.edu

Our talk at Lisbon SAMP conference was based mainly on our recent results on small diameter asymptotics for solutions of the Helmholtz equation in networks of thin fibers. These results were published in Ref. 21. The present paper contains a detailed review of Ref. 21 under some assumptions which make the results much more transparent. It also contains several new theorems on the structure of the spectrum near the threshold, small diameter asymptotics of the resolvent, and solutions of the evolution equation.

Keywords: Quantum graph, wave guide, Dirichlet problem spectrum, asymptotics

1. Introduction

This paper concerns the asymptotic spectral analysis of wave problems in systems of wave guides when the thickness ε of the wave guides vanishes. In the simplest case, the problem is described by the stationary wave (Helmholtz) equation

$$-\varepsilon^2 \Delta u = \lambda u, \quad x \in \Omega_\varepsilon, \quad Bu = 0 \quad \text{on } \partial \Omega_\varepsilon, \tag{1}$$

in a domain $\Omega_\varepsilon \subset R^d$, $d \geq 2$, with infinitely smooth boundary (for simplicity) and which has the following structure: Ω_ε is a union of a finite number of cylinders $C_{j,\varepsilon}$ (which we shall call channels) of lengths l_j, $1 \leq j \leq N$, with diameters of cross-sections of order $O(\varepsilon)$ and domains $J_{1,\varepsilon}, \cdots, J_{M,\varepsilon}$ (which we shall call junctions) connecting the channels into a network. It is assumed that the junctions have diameters of the same order $O(\varepsilon)$. The boundary condition has the form: $B = 1$ (the Dirichlet BC) or $B = \frac{\partial}{\partial n}$ (the Neumann BC) or $B = \varepsilon \frac{\partial}{\partial n} + \alpha(x)$, where n is the

---
[*]The authors were supported partially by the NSF grants DMS-0405927, DMS-0706928.

exterior normal and the function $\alpha > 0$ is real valued and does not depend on the longitudinal (parallel to the axis) coordinate on the boundary of the channels. One also can impose one type of BC on the lateral boundary of Ω_ε and another BC on free ends (which are not adjacent to a junction) of the channels. For simplicity we assume that only Dirichlet or Neumann BC are imposed on the free ends of the channels. Sometimes we shall denote the operator B on the lateral surface of Ω_ε by B_0, and we shall denote the operator B on the free ends of the channels by B_e.

Let m channels have infinite length. We start the numeration of $C_{j,\varepsilon}$ with the infinite channels. So, $l_j = \infty$ for $1 \leq j \leq m$. The axes of the channels form edges Γ_j, $1 \leq j \leq N$, of the limiting ($\varepsilon \to 0$) metric graph Γ. We split the set V of vertices v_j of the graph into two subsets $V = V_1 \cup V_2$, where the vertices from the set V_1 have degree 1 and the vertices from the set V_2 have degree at least two, i.e. vertices $v_i \in V_1$ of the graph Γ correspond to the free ends of the channels, and vertices $v_j \in V_2$ correspond to the junctions $J_{j,\varepsilon}$.

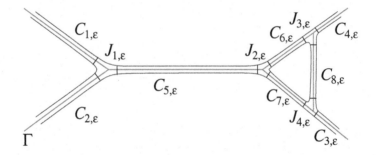

Fig. 1. An example of a domain Ω_ε with four junctions, four unbounded channels and four bounded channels.

Equation (1) degenerates when $\varepsilon = 0$. One could omit ε^2 in (1). However, the problem under consideration would remain singular, since the domain Ω_ε shrinks to the graph Γ as $\varepsilon \to 0$. The presence of this coefficient is convenient, since it makes the spectrum less vulnerable to changes in ε. As we shall see, in some important cases the spectrum of the problem does not depend on ε, and the spectrum will be magnified by a factor of ε^{-2} if ε^2 in (1) is omitted. The operator in $L^2(\Omega_\varepsilon)$ corresponding to the problem (1) will be denoted by H_ε.

The goal of this paper is the asymptotic analysis of the spectrum of H_ε, the resolvent $(H_\varepsilon - \lambda)^{-1}$, and solutions of the corresponding non-stationary problems for the heat and wave equations as $\varepsilon \to 0$. One can expect that H_ε is close (in some sense) to a one dimensional operator on the limiting graph Γ with appropriate gluing conditions at the vertices $v \in V$. The justification of this fact is not always simple. The form of the GC (gluing conditions) in the general situation was discovered quite recently in our previous paper.[21]

An important class of domains Ω_ε are the self-similar domains with only one junction and all the channels of infinite length. We shall call them *spider domains*. Thus, if Ω_ε is a spider domain, then there exists a point $\widehat{x} = x(\varepsilon)$ and an ε-independent domain Ω such that

$$\Omega_\varepsilon = \{(\widehat{x} + \varepsilon x) : x \in \Omega\}. \tag{2}$$

Thus, Ω_ε is the ε-contraction of $\Omega = \Omega_1$.

For the sake of simplicity we shall assume that Ω_ε is self-similar in a neighborhood of each junction. Namely, let $J_{j(v),\varepsilon}$ be the junction which corresponds to a vertex $v \in V$ of the limiting graph Γ. Consider a junction $J_{j(v),\varepsilon}$ and all the channels adjacent to $J_{j(v),\varepsilon}$. If some of these channels have finite length, we extend them to infinity. We assume that, for each $v \in V$, the resulting domain $\Omega_{v,\varepsilon}$ which consists of the junction $J_{j(v),\varepsilon}$ and the semi-infinite channels emanating from it is a spider domain. We also assume that all the channels $C_{j,\varepsilon}$ have the same cross-section ω_ε. This assumption is needed only to make the results more transparent. From the self-similarity assumption it follows that ω_ε is an ε-homothety of a bounded domain $\omega \subset R^{d-1}$.

Let $\lambda_0 < \lambda_1 \leq \lambda_2 ...$ be eigenvalues of the negative Laplacian $-\Delta_{d-1}$ in ω with the BC $B_0 u = 0$ on $\partial \omega$ where we put $\varepsilon = 1$ in B_0, and let $\{\varphi_n(y)\}$, $y \in \omega \in R^{d-1}$, be the set of corresponding orthonormal eigenfunctions. Then λ_n are eigenvalues of $-\varepsilon^2 \Delta_{d-1}$ in ω_ε and $\{\varphi_n(y/\varepsilon)\}$ are the corresponding eigenfunctions. In the presence of infinite channels, the spectrum of the operator H_ε consists of an absolutely continuous component which coincides with the semi-bounded interval $[\lambda_0, \infty)$ and a discrete set of eigenvalues. The eigenvalues can be located below λ_0 and can be embedded into the absolutely continuous spectrum. We will call the point $\lambda = \lambda_0$ the threshold since it is the bottom of the absolutely continuous spectrum or (and) the first point of accumulation of the eigenvalues as $\varepsilon \to 0$. Let us consider two of the simplest examples: the Dirichlet problem in a half infinite cylinder and in a bounded cylinder of the length l. In the first case, the spectrum of the negative Dirichlet Laplacian in Ω_ε is pure absolutely continuous and has multiplicity $n+1$ on the interval $[\lambda_n, \infty)$. In the second case the spectrum consists of the set of eigenvalues $\lambda_{n,m} = \lambda_n + \varepsilon^2 m^2/l^2$, $n \geq 0$, $m \geq 1$.

The wave propagation governed by the operator H_ε can be described in terms of the scattering solutions and scattering matrices associated with individual junctions of Ω_ε. The scattering solutions give information on the absolutely continuous spectrum and the resolvent for energies in the bulk of the spectrum ($\lambda > \lambda_0$). The spectrum in a small neighborhood of λ_0 and below λ_0 is associated with the parabolic equation. However, the scattering solutions allow us to approximate the operator H_ε, $\varepsilon \to 0$, by a one dimensional operator on the limiting graph for all values of $\lambda \geq \lambda_0$. In particular, when $\lambda \approx \lambda_0$ the corresponding GC on the limiting graph are expressed in terms of the limits of the scattering matrices as $\lambda \to \lambda_0$.

The plan of the paper is as follows. The next section is devoted to historical remarks. A more detailed description of the results from our paper[21] on the asymptotic behavior of the scattering solutions ($\lambda > \lambda_0 + \delta$, $\varepsilon \to 0$) is given in section 3. In particular, the GC on the limiting graph are described ($\lambda > \lambda_0 + \delta$). The Green function of the one dimensional problem on the limiting graph is studied in section 4. The resolvent convergence as $\varepsilon \to 0$ is established in section 5 when λ is near λ_0. This allows us to derive and rigorously justify the GC for the limiting problem with $\lambda \approx \lambda_0$ which were obtained earlier in Ref. 21 only on a formal level. They were understood as the limit of the GC with $\lambda > \lambda_0$ when $\lambda \to \lambda_0$. A detailed analysis of these GC is given.

Let $\lambda = \lambda_0 + O(\varepsilon^2)$. It has been known (see references in the next section) that for an arbitrary domain Ω_ε and the Neumann boundary condition on $\partial\Omega_\varepsilon$, the GC on the limiting graph Γ is Kirchhoff's condition. The GC are different for other boundary conditions on $\partial\Omega_\varepsilon$. It was shown in Ref. 21 that for generic domains Ω_ε and the boundary conditions different from the Neumann condition, the GC at the vertices of Γ are Dirichlet conditions. It is shown here that, for arbitrary domain Ω_ε, the GC at each vertex v of the limiting graph has the following form. For any function ς on Γ, we form a vector $\varsigma^{(v)}$ whose components are restrictions of ς to the edges of Γ adjacent to v. The GC at v, with λ near λ_0, are the Dirichlet condition for some components of the vector $\widetilde{\varsigma}^{(v)}$ and the Neumann condition for the remaining components where $\widetilde{\varsigma}^{(v)}$ is a rotation of $\varsigma^{(v)}$.

Note that the resolvent convergence provides the convergence of the discrete spectrum. In the presence of finite channels $C_{j,\varepsilon}$, the operator H_ε has a sequence of eigenvalues which converge to λ_0 as $\varepsilon \to 0$ (see the example above). Thus, these eigenvalues are asymptotically ($\varepsilon \to 0$) close to the eigenvalues of the problem where the junctions are replaced by Dirichlet/Neumann boundary conditions. The final result concerns the inverse scattering problem. The GC of the limiting problem depend on λ if $\lambda > \lambda_0 + \delta$. A λ-independent effective potential is constructed in the last section of the paper which has the same scattering data as the original problem. This allows one to reduce the problem in Ω_ε to a one dimensional problem with λ-independent GC.

2. Historical remarks

Certain problems related to the operator H_ε have been studied in detail. They concern, directly or indirectly, the spectrum near the origin for the operator H_ε with the Neumann boundary condition on $\partial\Omega_\varepsilon$, see Refs. 5,6,12, 13,15,18,19,25,26. The following couple of features distinguish the Neumann boundary condition. First, only in this case can the ground states $\varphi_0(y/\varepsilon) = 1$ on the cross sections of the channels be extended smoothly onto the junctions (by 1) to provide the ground state for the operator H_ε in an arbitrary domain Ω_ε. Another important fact, which is valid only in the case of the Neumann boundary conditions, is that $\lambda_0 = 0$. Note that an eigenvalue $\lambda = \mu$ of the operator H_ε contributes a term of order $e^{-\frac{\mu t}{\varepsilon^2}}$ to the solutions of the heat equation in Ω_ε. The existence of the spectrum in a small (of order $O(\varepsilon^2)$) neighborhood of the origin leads to the existence of a non-trivial limit, as $\varepsilon \to 0$, for the solutions of the heat equation. Solutions of the heat equation with other boundary conditions vanish exponentially as $\varepsilon \to 0$.

The GC and the justification of the limiting procedure $\varepsilon \to 0$ when λ is near $\lambda_0 = 0$ and the Neumann BC is imposed at the boundary of Ω_ε, can be found in Refs. 12,18,19,26. Typically, the GC at the vertices of the limiting graph in this case are: the continuity at each vertex v of both the field and the flow. These GC are called Kirchhoff's GC. The Ref. 12 provides the convergence, as $\varepsilon \to 0$, of the Markov process on Ω_ε to the Markov process on the limiting graph for more general domains Ω_ε (the cross section of a channel can vary). In the case when the shrinkage rate of the volume of the junctions is lower than the shrinkage rate of the area of the cross-sections of the guides, more complex, energy dependent or decoupling conditions may arise (see Refs. 5,15,19 for details).

The operator H_ε with the Dirichlet boundary condition on $\partial\Omega_\varepsilon$ was studied in a recent paper[24] under conditions that λ is near the threshold $\lambda_0 > 0$ and the junctions are more narrow than the channels. It is assumed there that the domain Ω_ε is bounded. Therefore, the spectrum of the operator (1) is discrete. It is proved that the eigenvalues of the operator (1) in a small neighborhood of λ_0 behave asymptotically, when $\varepsilon \to 0$, as eigenvalues of the problem in the disconnected domain that one gets by omitting the junctions, separating the channels in Ω_ε and adding the Dirichlet conditions on the bottoms of the channels. This result indicates that the waves do not propagate through the narrow junctions when λ is close to the bottom of the absolutely continuous spectrum. A similar result was obtained in Ref. 2 for the Schrödinger operator with a potential having a deep strict minimum on the graph, when the width of the walls shrinks to zero. It will be shown in this paper, that the same result (the GC on the limiting graph is the Dirichlet condition if H_ε is the operator with the Dirichlet boundary condition on $\partial\Omega_\varepsilon$ and $\lambda \approx \lambda_0 > 0$) is valid for generic domains Ω_ε without assumptions on the size of the junctions.

The asymptotic analysis of the scattering solutions and the resolvent for operator H_ε with arbitrary boundary conditions on $\partial\Omega_\varepsilon$ and λ in the bulk of the absolutely continuous spectrum ($\lambda > \lambda_0 + \delta$) was given by us in Refs. 20,21. It was shown there that the GC on the limiting graph can be expressed in terms of the scattering matrices defined by junctions of Ω_ε. Formal extension of these conditions to $\lambda = \lambda_0$ leads to the Dirichlet boundary conditions at the vertices of the limiting graph for generic domains Ω_ε. Among other results, we will show here that the asymptotics obtained in Refs. 20,21 are valid up to $\lambda = \lambda_0$.

There is extended literature on the spectrum of the operator H_ε below the threshold λ_0 (for example, see Refs. 3-11 and references therein). We shall not discuss this topic in the present paper. Important facts on the scattering solutions in networks of thin fibers can be found in Refs. 22,23.

3. Scattering solutions

We introduce Euclidean coordinates (t,y) in channels $C_{j,\varepsilon}$ chosen in such a way that the t-axis is parallel to the axis of the channel, hyperplane R_y^{d-1} is orthogonal to the axis, and $C_{j,\varepsilon}$ has the following form in the new coordinates:

$$C_{j,\varepsilon} = \{(t,\varepsilon y) : 0 < t < l_j,\ y \in \omega\}.$$

Let us recall the definition of scattering solutions for the problem in Ω_ε. In this paper, we'll need the scattering solutions only in the case of $\lambda \in (\lambda_0, \lambda_1)$. Consider the non-homogeneous problem

$$(-\varepsilon^2 \Delta - \lambda)u = f,\ x \in \Omega_\varepsilon;\quad Bu = 0\ \text{on}\ \partial\Omega_\varepsilon. \tag{3}$$

Definition 3.1. Let $f \in L^2_{com}(\Omega_\varepsilon)$ have a compact support, and $\lambda_0 < \lambda < \lambda_1$. A solution u of (3) is called outgoing if it has the following asymptotic behavior at infinity in each infinite channel $C_{j,\varepsilon}$, $1 \leq j \leq m$:

$$u = a_j e^{i\frac{\sqrt{\lambda-\lambda_0}}{\varepsilon}t}\varphi_0(y/\varepsilon) + O(e^{-\gamma t}), \quad \gamma = \gamma(\varepsilon,\lambda) > 0. \quad (4)$$

If $\lambda < \lambda_0$, a solution u of (3) is called outgoing if it decays at infinity.

Definition 3.2. Let $\lambda_0 < \lambda < \lambda_1$. A function $\Psi = \Psi_p^{(\varepsilon)}$, $0 \leq p \leq m$, is called a solution of the scattering problem in Ω_ε if

$$(-\varepsilon^2\Delta - \lambda)\Psi = 0, \quad x \in \Omega_\varepsilon; \quad B\Psi = 0 \text{ on } \partial\Omega_\varepsilon, \quad (5)$$

and Ψ has the following asymptotic behavior in each infinite channel $C_{j,\varepsilon}$, $1 \leq j \leq m$:

$$\Psi_p^{(\varepsilon)} = \delta_{p,j} e^{-i\frac{\sqrt{\lambda-\lambda_0}}{\varepsilon}t}\varphi_0(y/\varepsilon) + t_{p,j} e^{i\frac{\sqrt{\lambda-\lambda_0}}{\varepsilon}t}\varphi_0(y/\varepsilon) + O(e^{-\gamma t}), \quad (6)$$

where $\gamma = \gamma(\varepsilon,\lambda) > 0$, and $\delta_{p,j}$ is the Kronecker symbol, i.e. $\delta_{p,j} = 1$ if $p = j$, $\delta_{p,j} = 0$ if $p \neq j$.

The first term in (6) corresponds to the incident wave (coming through the channel $C_{p,\varepsilon}$), and all the other terms describe the transmitted waves. The transmission coefficients $t_{p,j} = t_{p,j}(\varepsilon,\lambda)$ depend on ε and λ. The matrix

$$T = [t_{p,j}] \quad (7)$$

is called the scattering matrix.

The outgoing and scattering solutions are defined similarly when $\lambda \in (\lambda_n, \lambda_{n+1})$. In this case, any outgoing solution has $n + 1$ waves in each channel propagating to infinity with the frequencies $\sqrt{\lambda - \lambda_s}/\varepsilon$, $0 \leq s \leq n$. There are $m(n + 1)$ scattering solutions: the incident wave may come through one of m infinite channels with one of $(n+1)$ possible frequencies. The scattering matrix has the size $m(n + 1) \times m(n + 1)$ in this case.

Theorem 3.1. *The scattering matrix T with $\lambda > \lambda_0$, $\lambda \notin \{\lambda_j\}$, is unitary and symmetric ($t_{p,j} = t_{j,p}$).*

The operator H_ε is non-negative, and therefore the resolvent

$$R_\lambda = (H_\varepsilon - \lambda)^{-1} : L^2(\Omega_\varepsilon) \to L^2(\Omega_\varepsilon) \quad (8)$$

is analytic in the complex λ plane outside the positive semi-axis $\lambda \geq 0$. Hence, the operator R_{k^2} is analytic in k in the half plane Im$k > 0$. We are going to consider the analytic extension of the operator R_{k^2} to the real axis and the lower half plane. Such an extension does not exist if R_{k^2} is considered as an operator in $L^2(\Omega_\varepsilon)$ since R_{k^2} is an unbounded operator when $\lambda = k^2$ belongs to the spectrum of the operator R_λ. However, one can extend R_{k^2} analytically if it is considered as an operator in the following spaces (with a smaller domain and a larger range):

$$R_{k^2} : L^2_{com}(\Omega_\varepsilon) \to L^2_{loc}(\Omega_\varepsilon). \tag{9}$$

Theorem 3.2.

(1) The spectrum of the operator H_ε consists of the absolutely continuous component $[\lambda_0, \infty)$ (if Ω_ε has at least one infinite channel) and, possibly, a discrete set of positive eigenvalues $\lambda = \mu_{j,\varepsilon}$ with the only possible limiting point at infinity. The multiplicity of the a.c. spectrum changes at points $\lambda = \lambda_n$, and is equal to $m(n+1)$ on the interval $(\lambda_n, \lambda_{n+1})$. If Ω_ε is a spider domain, then the eigenvalues $\mu_{j,\varepsilon} = \mu_j$ do not depend on ε.

(2) The operator (9) admits a meromorphic extension from the upper half plane $\mathrm{Im}\,k > 0$ into the lower half plane $\mathrm{Im}\,k < 0$ with the branch points at $k = \pm\sqrt{\lambda_n}$ of the second order and the real poles at $k = \pm\sqrt{\mu_{j,\varepsilon}}$ and, perhaps, at some of the branch points (see the remark below). The resolvent (9) has a pole at $k = \pm\sqrt{\lambda_n}$ if and only if the homogeneous problem (3) with $\lambda = \lambda_n$ has a nontrivial solution u such that

$$u = a_j \varphi_n(y/\varepsilon) + O(e^{-\gamma t}), \quad x \in C_{j,\varepsilon}, \quad t \to \infty, \quad 1 \leq j \leq m. \tag{10}$$

(3) If $f \in L^2_{com}(\Omega_\varepsilon)$, $\lambda > \lambda_0$, $k = \sqrt{\lambda}$ is real and is not a pole or a branch point of the operator (9), then the problem (3), (4) is uniquely solvable and the outgoing solution u can be found as the $L^2_{loc}(\Omega_\varepsilon)$ limit

$$u = R_{\lambda+i0} f. \tag{11}$$

(4) There exist exactly $m(n+1)$ different scattering solutions for the values of $\lambda \in (\lambda_n, \lambda_{n+1})$ such that $k = \sqrt{\lambda}$ is not a pole of the operator (9), and the scattering solution is defined uniquely after the incident wave is chosen.

Remark 3.1. The pole of R_λ at a branch point $\lambda = \lambda_n$ is defined as the pole of this operator function considered as a function of $z = \sqrt{\lambda - \lambda_n}$.

Let us describe the asymptotic behavior of scattering solutions $\Psi = \Psi_p^{(\varepsilon)}$ as $\varepsilon \to 0$, $\lambda \in (\lambda_0, \lambda_1)$. We shall consider here only the first zone of the absolutely continuous spectrum, but one can find the asymptotics of $\Psi_p^{(\varepsilon)}$ in Ref. 21 for any $\lambda > \lambda_0$. Note that an arbitrary solution u of equation (1) in a channel $C_{j,\varepsilon}$ can be represented as a series with respect to the orthogonal basis $\{\varphi_n(y/\varepsilon)\}$ of the eigenfunctions of the Laplacian in the cross-section of $C_{j,\varepsilon}$. Thus, it can be represented as a linear combination of the travelling waves

$$e^{\pm i \frac{\sqrt{\lambda - \lambda_n}}{\varepsilon} t} \varphi_n(y/\varepsilon), \quad \lambda \in (\lambda_n, \lambda_{n+1}),$$

and terms which grow or decay exponentially along the axis of $C_{j,\varepsilon}$. The main term of small ε asymptotics of the scattering solutions contains only travelling waves, i.e. functions $\Psi_p^{(\varepsilon)}$ in each channel $C_{j,\varepsilon}$ have the following form when $\lambda \in (\lambda_0, \lambda_1)$:

$$\Psi = \Psi_p^{(\varepsilon)} = \left(\alpha_{p,j} e^{-i \frac{\sqrt{\lambda - \lambda_0}}{\varepsilon} t} + \beta_{p,j} e^{i \frac{\sqrt{\lambda - \lambda_0}}{\varepsilon} t} \right) \varphi_0(y/\varepsilon) + r_{p,j}^\varepsilon, \quad x \in C_{j,\varepsilon}, \tag{12}$$

where

$$|r_{p,j}^\varepsilon| \le C e^{-\frac{\gamma d(t)}{\varepsilon}}, \quad \gamma > 0, \quad \text{and} \quad d(t) = \min(t, l_j - t).$$

The constants $\alpha_{p,j}$, $\beta_{p,j}$ and functions $r_{p,j}^\varepsilon$ depend on λ and ε. Formula (12) can be written as follows

$$\Psi = \Psi_p^{(\varepsilon)} = \varsigma \varphi_0(y/\varepsilon) + r_p^\varepsilon, \quad |r_p^\varepsilon| \le C e^{-\frac{\gamma d(t)}{\varepsilon}}, \tag{13}$$

where the function $\varsigma = \varsigma(t)$ can be considered as a function on the limiting graph Γ which is equal to $\varsigma_j(t)$, $0 < t < l_j$, on the edge Γ_j and satisfies the following equation:

$$\left(\varepsilon^2 \frac{d^2}{dt^2} + \lambda - \lambda_0 \right) \varsigma = 0. \tag{14}$$

In order to complete the description of the main term of the asymptotic expansion (12), we need to provide the choice of constants in the representation of ς_j as a linear combinations of the exponents. We specify ς by imposing conditions at infinity and gluing conditions (GC) at each vertex v of the graph Γ. Let $V = \{v\}$ be the set of vertices v of the limiting graph Γ. These vertices correspond to the free ends of the channels and the junctions in Ω_ε.

The conditions at infinity concern only the infinite channels $C_{j,\varepsilon}$, $j \le m$. They indicate that the incident wave comes through the channel $C_{p,\varepsilon}$. They have the form:

$$\beta_{p,j} = \delta_{p,j}. \tag{15}$$

The GC at vertices v of the graph Γ are universal for all incident waves and depend on λ. We split the set V of vertices v of the graph in two subsets $V = V_1 \cup V_2$, where the vertices from the set V_1 have degree 1 and correspond to the free ends of the channels, and the vertices from the set

V_2 have degree at least two and correspond to the junctions $J_{j,\varepsilon}$. We keep the same BC at $v \in V_1$ as at the free end of the corresponding channel of Ω_ε :

$$B_e\zeta = 0 \quad \text{at } v \in V_1. \tag{16}$$

In order to state the GC at a vertex $v \in V_2$, we choose the parametrization on Γ in such a way that $t = 0$ at v for all edges adjacent to this particular vertex. The origin ($t = 0$) on all the other edges can be chosen at any of the end points of the edge. Let $d = d(v) \geq 2$ be the order (the number of adjacent edges) of the vertex $v \in V_2$. For any function ς on Γ, we form a vector $\varsigma^{(v)} = \varsigma^{(v)}(t)$ with $d(v)$ components equal to the restrictions of ς on the edges of Γ adjacent to v. We shall need this vector only for small values of $t \geq 0$. Consider auxiliary scattering problems for the spider domain $\Omega_{v,\varepsilon}$. The domain is formed by the individual junction which corresponds to the vertex v, and all channels with an end at this junction, where the channels are extended to infinity if they have a finite length. We enumerate the channels of $\Omega_{v,\varepsilon}$ according to the order of the components of the vector $\varsigma^{(v)}$. We denote by Γ_v the limiting graph defined by $\Omega_{v,\varepsilon}$. Definitions 3.1, 3.2 and Theorem 3.2 remain valid for the domain $\Omega_{v,\varepsilon}$. In particular, one can define the scattering matrix $T = T_v(\lambda)$ for the problem (1) in the domain $\Omega_{v,\varepsilon}$. Let I_v be the unit matrix of the same size as the size of the matrix $D_v(\lambda)$. The GC at the vertex $v \in V_2$ has the form

$$i\varepsilon[I_v + T_v(\lambda)]\frac{d}{dt}\varsigma^{(v)}(t) - \sqrt{\lambda - \lambda_0}[I_v - T_v(\lambda)]\varsigma^{(v)}(t) = 0, \quad t = 0, \quad v \in V_2. \tag{17}$$

One has to keep in mind that the self-similarity of the spider domain $\Omega_{v,\varepsilon}$ implies that $T_v = T_v(\lambda)$ does not depend on ε.

Definition 3.3. *A family of subsets $l(\varepsilon)$ of a bounded closed interval $l \subset R^1$ will be called thin if, for any $\delta > 0$, there exist constants $\beta > 0$ and c_1, independent of δ and ε, and $c_2 = c_2(\delta)$, such that $l(\varepsilon)$ can be covered by c_1 intervals of length δ together with $c_2\varepsilon^{-1}$ intervals of length $c_2 e^{-\beta/\varepsilon}$. Note that $|l(\varepsilon)| \to 0$ as $\varepsilon \to 0$.*

Theorem 3.3. *For any bounded closed interval $l \subset (\lambda_0, \lambda_1)$, there exists $\gamma = \gamma(l) > 0$ and a thin family of sets $l(\varepsilon)$ such that the asymptotic expansion (13) holds on all (finite and infinite) channels $C_{j,\varepsilon}$ uniformly in $\lambda \in l \setminus l(\varepsilon)$ and x in any bounded region of R^d. The function ς in (13) is a vector function on the limiting graph which satisfies the equation (14), conditions (15) at infinity, BC (16), and the GC (17).*

Remark 3.2.

(1) For spider domains, the estimate of the remainder is uniform for all $x \in R^d$.

(2) The asymptotics stated in Theorem 3.3 is valid only outside of a thin set $l(\varepsilon)$ since the poles of the resolvent (8) may run over the interval (λ_0, λ_1) as $\varepsilon \to 0$, and the scattering solution may not exist when λ is a pole of the resolvent. These poles do not depend on ε for spider domains, and the set $l(\varepsilon)$ is ε-independent in this case.

Consider a spider domain $\Omega_{v,\varepsilon}$ and scattering solutions $\Psi = \Psi_p^{(\varepsilon)}$ in $\Omega_{v,\varepsilon}$ when λ belongs to a small neighborhood of λ_0, i.e.

$$\Psi = \Psi_{p,v}^{(\varepsilon)} = \left(\delta_{p,j} e^{-i\frac{\sqrt{\lambda-\lambda_0}}{\varepsilon}t} + t_{p,j}(\lambda) e^{i\frac{\sqrt{\lambda-\lambda_0}}{\varepsilon}t}\right) \varphi_0(y/\varepsilon) + r_{p,j}^\varepsilon,$$

$$x \in C_{j,\varepsilon}, \quad |\lambda - \lambda_0| \leq \delta, \quad (18)$$

where $r_{p,j}^\varepsilon$ decays exponentially as $t \to \infty$ and $t = t(x)$ is the coordinate of the point x. We define these solutions for all complex λ in the circle $|\lambda - \lambda_0| \leq \delta$ by the asymptotic expansion (18) when $\text{Im}\lambda \geq 0, \lambda \neq \lambda_0$, and by extending them analytically for other values of λ in the circle.

Lemma 3.1. *Let $\Omega_{v,\varepsilon}$ be a spider domain. Then there exist δ and $\gamma > 0$ such that*

(1) for each p and $|\lambda - \lambda_0| \leq \delta$, $\lambda \neq \lambda_0$, the scattering solution exists and is unique,

(2) the scattering coefficients $t_{p,j}(\lambda)$ are analytic in $\sqrt{\lambda - \lambda_0}$ when $|\lambda - \lambda_0| \leq \delta$,

(3) the following estimate is valid for the remainder

$$|r_{p,j}^\varepsilon| \leq \frac{C}{|\lambda - \lambda_0|} e^{-\frac{\gamma t}{\varepsilon}}, \quad \gamma > 0, \quad x \in C_{j,\varepsilon}.$$

This statement can be extracted from our paper.[21] Since it was not stated explicitly, we shall derive it from the theorems above. In fact, since the spider domain $\Omega_{v,\varepsilon}$ is self-similar, it is enough to prove this lemma when $\varepsilon = 1$. We omit index ε in Ω_ε. $C_{p,\varepsilon}, \Psi_p^{(\varepsilon)}$ when the problem in Ω_ε is considered with $\varepsilon = 1$. Let $\alpha_p(x)$ be a C^∞- function on Ω which is equal to zero outside of the channel C_p and equal to one on C_p when $t > 1$. We look for the solution Ψ_p of the scattering problem in the form

$$\Psi_p = \delta_{p,j} e^{-i\sqrt{\lambda-\lambda_0}t}\varphi_0(y)\alpha_p(x) + u_p. \quad \Im\lambda \geq 0, \lambda \neq \lambda_0,$$

Then u_p is the outgoing solution of the problem

$$(-\Delta - \lambda)u = f, \quad x \in \Omega; \quad u = 0 \text{ on } \partial\Omega,$$

where

$$f = -\delta_{p,j}\left[2\nabla\left(e^{i\sqrt{\lambda-\lambda_0}t}\varphi_0(y)\right)\nabla\alpha_p(x) + e^{i\sqrt{\lambda-\lambda_0}t}\varphi_0(y)\Delta\alpha_p(x)\right] \in L^2_{com}(\Omega).$$

From Theorem 3.2 it follows that there exists $\delta > 0$ such that u_p exists and is unique when $|\lambda - \lambda_0| \leq \delta$, $\Im\lambda \geq 0$, $\lambda \neq \lambda_0$, and $u_p = R_\lambda f$ when $\Im\lambda \geq 0$, and u_p can be extended analytically to the lower half-plane if R_λ is understood as in (9). The function $u_p = R_\lambda f$ may have a pole at $\lambda = \lambda_0$. In particular, on the cross-sections $t = 1$ of the infinite channels C_j, the function u_p is analytic in $\sqrt{\lambda - \lambda_0}$ (when $|\lambda - \lambda_0| \leq \delta$) with a possible pole at $\lambda = \lambda_0$. We note that $f = 0$ on $C_j \cap \{t \geq 1\}$. We represent u_p there as a series with respect to the basis $\{\varphi_s(y)\}$. This leads to (18) and justifies all the statements of the lemma if we take into account the following two facts: 1) the resolvent (9) cannot have a singularity at $\lambda = \lambda_0$ of order higher than $1/|\lambda-\lambda_0|$, since the norm of the resolvent (8) at any point λ does not exceed the inverse distance from λ to the spectrum, 2) the scattering coefficients cannot have a singularity at $\lambda = \lambda_0$ due to Theorem 3.1.

The proof of Lemma 3.1 is complete.

4. Spectrum of the problem on the limiting graph

Let us write the inhomogeneous problem on the limiting graph Γ which corresponds to the scattering problem (14), (15), (16), (17). We shall always assume that the function f in the right-hand side of the equation below has compact support. Then the corresponding inhomogeneous problem has the form

$$\left(\varepsilon^2\frac{d^2}{dt^2} + \lambda - \lambda_0\right)\varsigma = f \quad \text{on } \Gamma, \qquad (19)$$

$B_e\varsigma = 0$ at $v \in V_1$,

$$i\varepsilon[I_v + T_v(\lambda)]\frac{d}{dt}\varsigma^{(v)}(0) - \sqrt{\lambda - \lambda_0}[I_v - T_v(\lambda)]\varsigma^{(v)}(0) = 0 \quad \text{at } v \in V_2,$$
(20)

$$\varsigma = \beta_j e^{i\frac{\sqrt{\lambda-\lambda_0}}{\varepsilon}t}, \quad t \gg 1, \quad \text{on the infinite edges } \Gamma_j, \ 1 \leq j \leq m. \quad (21)$$

This problem is relevant to the original problem in Ω_ε only while $\lambda < \lambda_1$, since more than one mode in each channel survives as $\varepsilon \to 0$ when $\lambda > \lambda_1$.

The latter leads to a more complicated problem on the limiting graph (see Ref. 21). We are going to use the problem (19)-(21) to study the spectrum of the operator H_ε when

$$\lambda = \lambda_0 + \varepsilon^2 \mu, \quad |\mu| < c. \tag{22}$$

As we shall see later, if Ω_ε has a channel of finite length, then the operator H_ε has a sequence of eigenvalues which are at a distance of order $O(\varepsilon^2)$ from the threshold $\lambda = \lambda_0$. For example, if Ω_ε is a finite cylinder with the Dirichlet boundary condition (see the introduction) these eigenvalues have the form $\lambda_0 + \varepsilon^2 m^2 / l^2$, $m \geq 1$. Thus, assumption (22) allows one to study any finite number of eigenvalues near $\lambda = \lambda_0$.

Let us make the substitution $\lambda = \lambda_0 + \varepsilon^2 \mu$ in (19)-(21). Condition (20) may degenerate at $\mu = 0$, and one needs to understand this condition at $\mu = 0$ as the limit when $\mu \to 0$ after an appropriate normalization which will be discussed later.

Lemma 4.1. *There is an orthogonal projection $P = P(v)$ in $R^{d(v)}$ such that the problem (19)-(21) under condition (22) can be written in the form*

$$\left(\frac{d^2}{dt^2} + \mu\right)\varsigma = \varepsilon^{-2} f \quad \text{on } \Gamma; \tag{23}$$

$B_e \varsigma = 0$ at $v \in V_1$;

$$P\varsigma^{(v)}(0) + O(\varepsilon)\frac{d}{dt}\varsigma^{(v)}(0) = 0, \; P^\perp \frac{d}{dt}\varsigma^{(v)}(0) + O(\varepsilon)\varsigma^{(v)}(0) = 0 \;\; at \; v \in V_2,$$
$$\tag{24}$$

$$\varsigma = \beta_j e^{i\sqrt{\mu}t}, \quad t \gg 1, \quad \text{on the infinite edges } \Gamma_j, \; 1 \leq j \leq m, \tag{25}$$

where $d(v)$ is the order (the number of adjacent edges) of the vertex v and $P^\perp$ is the orthogonal complement to P.

Remark 4.1.

(1) The GC (23) at $v \in V_2$ looks particularly simple in the eigenbasis of the operators P, $P^\perp$. If $\varepsilon = 0$, then it is the Dirichlet/Neumann GC, i.e after an appropriate orthogonal transformation $\xi^v = C_v \tilde{\xi}^v$,

$$\tilde{\xi}_1^v(0) = \cdots = \tilde{\xi}_k^v(0) = 0, \; \frac{d\tilde{\xi}_{k+1}^v}{dt}(0) = \cdots = \frac{d\tilde{\xi}_d^v}{dt}(0) = 0, \; k = \text{Rank} P.$$

(2) We consider μ as being a spectral parameter of the problem (23)-(25), but one needs to keep in mind that the terms $O(\varepsilon)$ in condition (23) depend on μ.

Proof. Let us recall that the matrix $T_v(\lambda)$ is analytic in $\sqrt{\lambda - \lambda_0}$ due to Lemma 3.1. Theorem 3.1 implies the existence of the orthogonal matrix C_v such that $D_v := C_v^{-1} T(\lambda_0) C_v$ is a diagonal matrix with elements $\nu_s = \pm 1$, $1 \leq s \leq d(v)$, on the diagonal. In fact, from Theorem 3.1 it follows that, for any $\lambda \in [\lambda_0, \lambda_1]$, one can reduce $T(\lambda)$ to a diagonal form with diagonal elements $\nu_s = \nu_s(\lambda)$ where $|\nu_s| = 1$. Additionally, one can easily show that the matrix $T(\lambda_0)$ is real-valued, and therefore, $\nu_s = \pm 1$ when $\lambda = \lambda_0$. The statement of the lemma follows immediately from here with $P = \frac{1}{2}(I - D_v)C_v^{-1}$. This completes the proof. $\square$

Consider the Green function $G_\varepsilon(\gamma, \gamma_0, \mu)$, $\gamma, \gamma_0 \in \Gamma$, $\gamma_0 \notin V$, of the problem (23)-(25) which is the solution of the problem with $\varepsilon^{-2} f$ replaced by $\delta_{\gamma_0}(\gamma)$. Here $\delta_{\gamma_0}(\gamma)$ is the delta function on Γ supported at the point γ_0 which belongs to one of the edges of Γ (γ_0 is not a vertex). The Green function $G_0(\gamma, \gamma_0, \mu)$ is the solution of (23)-(25) with $\varepsilon = 0$ in (24).

Let us denote by P_0 the closure in $L^2(\Gamma)$ of the operator $-\frac{d^2}{dt^2}$ defined on smooth functions satisfying (24) with $\varepsilon = 0$. Note that the conditions (24) with $\varepsilon = 0$ do not depend on μ. Hence, P_0 is a self-adjoint operator whose spectrum consists of an absolutely continuous component $\{\mu \geq 0\}$ (if Γ has at least one unbounded edge) and a discrete set $\{\mu_j\}$ of non-negative eigenvalues. Let us denote by B_c the disk $|\mu| < c$ of the complex μ-plane.

Lemma 4.2. *For any $c > 0$ there exist $\varepsilon_0 > 0$ such that*

(1) the eigenvalues $\{\mu_s(\varepsilon)\}$ of the problem (23)-(25) in the disk B_c of the complex μ-plane are located in $C_1\varepsilon$-neighborhoods of the points $\{\mu_j\}$, $C_1 = C_1(c)$, and each such neighborhood contains p eigenvalues $\mu_s(\varepsilon)$ with multiplicity taken into account, where p is the multiplicity of the eigenvalue μ_j.

(2) the Green function G_ε exists and is unique when $\mu \in B_c \setminus \{\mu_s(\varepsilon)\}$ and has the form

$$G_\varepsilon(\gamma, \gamma_0, \mu) = \frac{g(\gamma, \gamma_0, \mu, \varepsilon)}{h(\mu, \varepsilon)},$$

where g is a continuous function of $\gamma \in \Gamma_m$, $\gamma_0 \in \Gamma_n$, $\mu \in B_c$, $\varepsilon \in [0, \varepsilon_0]$, functions g and h are analytic in $\sqrt{\mu}$ and ε, and h has zeros in B_c only at points $\mu = \mu_s(\varepsilon)$. Here Γ_m, Γ_n are arbitrary edges of Γ.

Proof. We denote by $t = t(\gamma)$ ($t_0 = t_0(\gamma_0)$) the value of the parameter on the edge of Γ which corresponds to the point $\gamma \in \Gamma_m$ ($\gamma_0 \in \Gamma_n$, respectively).

Laplace operator in networks of thin fibers: Spectrum near the threshold   83

We look for the Green function in the form

$$G_\varepsilon = \delta_{m,n}\frac{e^{i\sqrt{\mu}|t-t_0|}}{2i\sqrt{\mu}} + a_{m,n}e^{i\sqrt{\mu}t} + b_{m,n}e^{-i\sqrt{\mu}t}, \quad \gamma \in \Gamma_m, \ \gamma_0 \in \Gamma_n, \quad (26)$$

where $\delta_{m,n}$ is the Kronecker symbol, and the functions $a_{m,n}$, $b_{m,n}$ depend on t_0, μ, ε. Obviously, (23) with $\varepsilon^{-2}f$ replaced by $\delta_{\gamma_0}(\gamma)$ holds. Let us fix the edge Γ_n which contains γ_0. We substitute (26) into (24), (25) and get $2N$ equations for $2N$ unknowns $a_{m,n}$, $b_{m,n}$, $1 \leq m \leq N$, where n is fixed. The matrix M of this system depends analytically on $\sqrt{\mu}$, $\mu \in B_c$ and $\varepsilon \in [0, \varepsilon_0]$. The right-hand side has the form $c_1 e^{i\sqrt{\mu}t_0} + c_2 e^{-i\sqrt{\mu}t_0}$, where the vectors c_1, c_2 depend analytically on $\sqrt{\mu}$ and ε. This implies all the statements of the lemma if we take into account that the determinant of M with $\varepsilon = 0$ has zeroes at eigenvalues of the operator P_0. The proof is complete. □

In order to justify the resolvent convergence of the operator H_ε as $\varepsilon \to 0$ and obtain the asymptotic behavior of the eigenvalues of the problem (1) near λ_0 we need to represent the Green function G_ε of the problem on the graph in a special form. We fix points γ_i strictly inside of the edges Γ_i. These points split Γ into graphs Γ_v^{cut} which consist of one vertex v and parts of adjacent edges up to corresponding points γ_i. If Γ_v is the limiting graph which corresponds to the spider domain $\Omega_{v,\varepsilon}$, then Γ_v^{cut} is obtained from Γ_v by cutting its edges at points γ_i.

When $\varepsilon \geq 0$ is small enough, equation (23) on Γ_v has $d(v)$ linearly independent solutions satisfying the condition from (24) which corresponds to the chosen vertex v. This is obvious if $\varepsilon = 0$ (when the components of the vector $C_v^{-1}\varsigma^{(v)}$ satisfy either the Dirichlet or the Neumann conditions at v). Therefore it is also true for small $\varepsilon \geq 0$. We denote this solution space by S_v. Let us fix a specific basis $\{\psi_{p,v}(\gamma,\mu,\varepsilon)\}$, $1 \leq p \leq d(v)$, in S_v. It is defined as follows. Let us change the numeration of the edges of Γ (if needed) in such a way that the first $d(v)$ edges are adjacent to v. We also choose the parametrization on these edges in such a way that $t = 0$ corresponds to v. Then

$$\psi_{p,v} = \delta_{p,j}e^{-i\sqrt{\mu}t} + t_{p,j}(\lambda)e^{i\sqrt{\mu}t}, \quad \gamma \in \Gamma_j.$$

Here $t = t(\gamma)$, $\lambda = \lambda_0 + \varepsilon^2\mu$, $t_{p,j} = t_{p,j}^{(v)}$ are the scattering coefficients for the spider domain $\Omega_{v,\varepsilon}$. Obviously, $\psi_{p,v}$ satisfies conditions (24), and formula (18) can be written as

$$\Psi_{p,v}^{(\varepsilon)} = \psi_{p,v}(\gamma,\mu,\varepsilon)\varphi_0(y/\varepsilon) + r_{p,j}^\varepsilon, \quad x \in C_{j,\varepsilon}, \quad \gamma = \gamma(x). \quad (27)$$

where $\gamma(x) \in \Gamma_j$ is defined by the cross-section of the channel $C_{j,\varepsilon}$ through the point x.

We shall choose one of the points γ_j in a special way. Namely, if $\gamma_0 \in \Gamma_n$ then we chose $\gamma_n = \gamma_0$. Then G_ε belongs to the solution space S_v, and from Lemma 4.2 we get

Lemma 4.3. *The Green function G_ε can be represented on each part Γ_v^{cut} of the graph Γ in the form*

$$G_\varepsilon(\gamma, \gamma_0, \mu) = \frac{1}{h(\mu, \varepsilon)} \sum_{1 \leq p \leq d(v)} a_{p,v} \psi_{p,v}(\gamma, \mu, \varepsilon), \quad \mu \in B_c, \ \varepsilon \in [0, \varepsilon_0], \tag{28}$$

where the function h is defined in Lemma 4.2 and $a_{p,v} = a_{p,v}(\gamma_0, \mu, \varepsilon)$ are continuous functions which are analytic in ε and $\sqrt{\mu}$

5. Resolvent convergence of the operator H_ε

We are going to study the asymptotic behavior of the resolvent $R_{\lambda,\varepsilon} = (H_\varepsilon - \lambda)^{-1}$ of the operator H_ε when (22) holds and $\varepsilon \to 0$. When μ is complex, the resolvent R_λ is understood in the sense of analytic continuation described in Theorem 3.2. In fact, we shall study $R_\lambda f$ only inside of the channels $C_{j,\varepsilon}$ and under the assumption that the support of f belongs to a bounded region inside of the channels. We fix finite segments $\Gamma'_j \subset \Gamma_j$ of the edges of the graph large enough to contain the points γ_j. Let $\Gamma' = \cup \Gamma'_j$. We denote by $C'_\varepsilon = \cup C'_{j,\varepsilon}$ the union of the finite parts $C'_{j,\varepsilon}$ of the channels which shrink to Γ'_j as $\varepsilon \to 0$. We shall identify functions from $L^2(C'_\varepsilon)$ with functions from $L^2(\Omega_\varepsilon)$ equal to zero outside C'_ε. We also omit the restriction operator when functions on Ω_ε are considered only on C'_ε.

If $f \in L^2(C'_\varepsilon)$, denote

$$\widehat{f}(\gamma, \varepsilon) = \frac{<f, \varphi_0(y/\varepsilon)>}{\|\varphi_0(y/\varepsilon)\|_{L^2}} = \int_{\Omega_\varepsilon} f \varphi_0(y/\varepsilon) dy / \|\varphi_0(y/\varepsilon)\|_{L^2}, \quad \gamma \in \Gamma.$$

We shall use the notation G_ε for the integral operator

$$G_\varepsilon \zeta(\gamma) = \int_\Gamma G_\varepsilon(\gamma, \gamma_0, \mu) \zeta(\gamma_0) d\gamma_0, \quad \gamma \in \Gamma.$$

Theorem 5.1. *Let (22) hold. Then for any disk B_c, there exist $\varepsilon_0 = \varepsilon_0(c)$ and a constant $C < \infty$ such that the function*

$$R_{\lambda,,\varepsilon} f = (H_\varepsilon - \lambda)^{-1} f, \quad f \in L^2(C'_\varepsilon),$$

is analytic in $\sqrt{\mu}$ when $\mu \in B_c \backslash O(\varepsilon)$, where $O(\varepsilon)$ is $C\varepsilon$-neighborhood of the set $\{\mu_j\}$, and has the form

$$\|R_{\lambda,\varepsilon} f - \varphi_0(y/\varepsilon) G_\varepsilon \widehat{f}(\gamma,\varepsilon)\|_{L^2(C'_\varepsilon)} \leq C\varepsilon \|f\|_{L^2(C'_\varepsilon)}.$$

Remark 5.1.

(1) The points μ_j were introduced above as eigenvalues of the problem (23)-(25) on the graph with $\varepsilon = 0$. The GC in this case are the Dirichlet and Neumann conditions for the components of the vector $C_v^{-1} \varsigma^{(v)}$. Obviously, these points are also eigenvalues of the operator H_ε with the junctions of Ω_ε replaced by the same Dirichlet/Neumann conditions on the edges of the channels adjacent to the junctions.

(2) The resolvent convergence stated in the theorem implies the convergence, as $\varepsilon \to 0$, of eigenvalues of operator H_ε to $\{\mu_j\}$. We could not guarantee the fact that the eigenvalues of the problem on the graph are real (see Lemma 4.2). Of course, they are real for operator H_ε.

Proof. We construct an approximation $K_{\lambda,\varepsilon}$ to the resolvent $R_{\lambda,\varepsilon} = (H_\varepsilon - \lambda)^{-1}$ for $f \in L^2(C'_\varepsilon)$. We represent $L^2(C'_\varepsilon)$ as the orthogonal sum

$$L^2(C'_\varepsilon) = L_0^2(C'_\varepsilon) + L_1^2(C'_\varepsilon),$$

where functions from $L_0^2(C'_\varepsilon)$ have the form $h(\gamma)\varphi_0(y/\varepsilon)$, $\gamma \in \Gamma'$, and functions from $L_1^2(C'_\varepsilon)$ on each cross-section of the channels are orthogonal to $\varphi_0(y/\varepsilon)$. Here and below the point $\gamma = \gamma(x) \in \Gamma$ is defined by the cross-section of the channel through x. We put $\gamma_0 = \gamma(x_0)$, i.e. γ_0 is the point on the graph defined by the cross-section of the channel through x_0.

Consider the operator

$$K_{\lambda,\varepsilon} : L^2(C'_\varepsilon) \to L^2(\Omega_\varepsilon)$$

with kernel $K_{\lambda,\varepsilon}(x,x_0)$ defined as follows:

$$K_{\lambda,\varepsilon}(x,x_0) = \sum_{v \in V} \frac{1}{h(\varepsilon,\mu)} \sum_{1 \leq p \leq d(v)} a_{p,v} \widehat{\Psi}_{p,v}^{(\varepsilon)}(x,x_0), \qquad x_0 \in C'_\varepsilon, \; x \in \Omega_\varepsilon.$$

Here $h(\mu,\varepsilon)$ and $a_{p,v} = a_{p,v}(\gamma_0,\mu,\varepsilon)$ are the functions defined in (28). The functions $\widehat{\Psi}_{p,v}^{(\varepsilon)}$ are defined by the scattering solutions $\Psi_{p,v}^{(\varepsilon)}$ of the problem in the spider domain $\Omega_{v,\varepsilon}$ in the following way. Let $\Omega_{v,\varepsilon}^0$ be the part of the spider domain $\Omega_{v,\varepsilon}$ which consists of the junction and parts of the adjacent channels up to the cylinders $C'_{j,\varepsilon}$. Let $\Omega'_{v,\varepsilon}$ ($\Omega^1_{v,\varepsilon}$) be a bigger domain which contains additionally the parts of the cylinders $C'_{j,\varepsilon}$ up to the cross-sections which correspond to points γ_j (the whole cylinders $C'_{j,\varepsilon}$, respectively). We put $\widehat{\Psi}_{p,v}^{(\varepsilon)} = \Psi_{p,v}^{(\varepsilon)}$ in $\Omega_{v,\varepsilon}^0$. We split the scattering solutions $\Psi_{p,v}^{(\varepsilon)}$ in the cylinders $C_{j,\varepsilon}'$ into the sum of two terms. The first term contains the main modes $\varphi_0(y/\varepsilon)e^{\pm i\sqrt{\mu}t}$, and the second one is orthogonal to $\varphi_0(y/\varepsilon)$ in each cross-section. We multiply the first term by the function $\theta_v(x, x_0)$ equal to one on $\Omega_{v,\varepsilon}'$ and equal to zero everywhere else on Ω_ε. We multiply the second term by an infinitely smooth cut-off function $\eta_v(x)$ equal to one on $\Omega_{v,\varepsilon}^0$ and equal to zero on Ω_ε outside $\Omega_{v,\varepsilon}^1$. In other terms,

$$\widehat{\Psi}_{p,v}^{(\varepsilon)}(x, x_0) = \theta_v(x, x_0)\Psi_{p,v}^{(\varepsilon)} + (\eta_v(x) - \theta_v(x, x_0))r_{p,j}^\varepsilon, \qquad (29)$$

where $r_{p,j}^\varepsilon$ is defined in (27).

Recall that the representation $\Gamma = \cup \Gamma_v^{cut}$ depends on the choice of points $\gamma_s \in \Gamma_s' \subset \Gamma_s$. All these points are fixed arbitrarily except one: if $\gamma_0 \in \Gamma_j$ then γ_j is chosen to be equal to γ_0. This is the reason why θ_v depends on x_0 and η_v is x_0-independent.

Let (22) hold and $f \in L^2(C_\varepsilon')$. We look for the parametrix (almost resolvent) of the operator H_ε in the form

$$F_{\lambda,\varepsilon} = K_{\lambda,\varepsilon}P_0 + R_{\lambda,\varepsilon}P_1,$$

where P_0, P_1 are projections on $L_0^2(C_\varepsilon')$ and $L_1^2(C_\varepsilon')$, respectively. It is not difficult to show that $||R_{\lambda,\varepsilon}P_1|| = O(\varepsilon^2)$ and $H_\varepsilon F_{\lambda,\varepsilon} = I + F_{\lambda,\varepsilon}$, where $||F_{\lambda,\varepsilon}|| = O(\varepsilon)$. This implies that $R_{\lambda,\varepsilon} = K_{\lambda,\varepsilon}P_0 + O(\varepsilon)$. The latter, together with Lemma 3.1, justifies Theorem 5.1. □

6. The GC at λ near the threshold λ_0

Theorem 5.1 and the remarks following the theorem indicate that the GC at each vertex when $\lambda - \lambda_0 = O(\varepsilon^2)$ is the Dirichlet/Neumann condition, i.e. the junctions of Ω_ε can be replaced by $k(v)$ Dirichlet and $d(v) - k(v)$ Neumann conditions at the edges of the channels adjacent to the junctions (after an appropriate orthogonal transformation). We are going to specify the choice between the Dirichlet and Neumann conditions. First, we would like to make four important remarks.

Remark 6.1.

(1) Classical Kirchhoff's GC corresponds to $k = d - 1$.
(2) For any domain Ω_ε under consideration, if $\lambda - \lambda_0 = O(\varepsilon^2)$ and the Neumann boundary condition is imposed on $\partial\Omega_\varepsilon$ ($\lambda_0 = 0$ in this case) then the GC on the limiting graph is Kirchhoff's condition (see section 2).

(3) It was proven in Ref. 21 that if $\lambda - \lambda_0 = O(\varepsilon^2)$ and the boundary condition on $\partial\Omega_\varepsilon$ is different from the Neumann condition, then the GC on the limiting graph is the Dirichlet condition ($k = d$) for generic domains Ω_ε. An example at the end of the next section illustrates this fact.

(4) The theorem below states that Kirchhoff's GC condition on the limiting graph appears in the case of arbitrary boundary conditions on $\partial\Omega_\varepsilon$, if the operator H_ε has a ground state at $\lambda = \lambda_0$. The ground state at $\lambda = \lambda_0 = 0$ exists for an arbitrary domain Ω_ε, if the Neumann boundary condition is imposed on $\partial\Omega_\varepsilon$. The ground state at $\lambda = \lambda_0$ does not exist for generic domains Ω_ε in the case of other boundary conditions (see Ref. 21).

Note that the GC is determined by the scattering matrix in the spider domain $\Omega_{v,\varepsilon}$, and this matrix does not depend on ε. Thus, when the GC is studied, it is enough to consider a spider ε-independent domain $\Omega = \Omega_{v,1}$. We shall omit the indices v and ε in H_ε, $C_{j,\varepsilon}$ when $\varepsilon = 1$.

Definition 6.1. A ground state of the operator H in a spider domain Ω at $\lambda = \lambda_0$ is the function $\psi_0 = \psi_0(x)$, which is bounded, strictly positive inside Ω, satisfies the equation $(-\Delta - \lambda_0)\psi_0 = 0$ in Ω, and the boundary condition on $\partial\Omega$, and has the following asymptotic behavior at infinity

$$\psi_0(x) = \varphi_0(y)[\rho_j + o(1)], \quad x \in C_j, \; |x| \to +\infty, \tag{30}$$

where $\rho_j > 0$ and φ_0 is the ground state of the operator in the cross-sections of the channels.

Let us stress that we assume the strict positivity of ρ_j.

Let's consider the parabolic problem in a spider domain Ω_ε,

$$\frac{\partial u_\varepsilon}{\partial \tau} = \Delta u_\varepsilon, \quad u_\varepsilon(0, x) = \varphi(\gamma)\varphi_0(\frac{y}{\varepsilon}), \quad u_\varepsilon(\tau, x)|_{\partial\Omega_\varepsilon} = 0, \tag{31}$$

where $\gamma = \gamma(x) \in \Gamma$ is defined by the cross-section of the channel through the point x, function φ is continuous, compactly supported with a support outside of the junctions, and depends only on the longitudinal ("slow") variable t on each edge $\Gamma_j \subset \Gamma$. We shall denote the coordinate t on C_j and Γ_j by t_j. Let ω' be a compact in the cross-section ω of the channels C_j.

Theorem 6.1. *Let Ω be a spider domain, the Dirichlet or Robin boundary condition be imposed at $\partial\Omega$, and let the operator H have a ground state at $\lambda = \lambda_0$. Then asymptotically, as $\varepsilon \to 0$, the solution of the parabolic problem (31) in Ω_ε has the following form*

$$u_\varepsilon(\tau, x) = e^{-\frac{\lambda_0 \tau}{\varepsilon^2}} w_\varepsilon(\tau, x)\varphi_0\left(\frac{y}{\varepsilon}\right),$$

where the function w_ε converges uniformly in any region of the form $0 < c^{-1} < \tau < c$, $t_j(x) > \delta > 0$, $y \in \omega'$ to a function $w(\tau,\gamma)$ on the limiting graph Γ which satisfies the relations

$$\frac{\partial w}{\partial \tau} = \frac{\partial^2 w}{\partial t^2} \quad \text{on } \Gamma; \quad w \text{ is continuous at the vertex}, \quad \sum_{j=1}^{d} \rho_j \frac{\partial w}{\partial t_j}(0) = 0.$$
(32)

Remark 6.2.

(1) Let's note that under the ground state condition, the operator H_ε has no eigenvalues below λ_0. Otherwise, the eigenfunction with the eigenvalue $\lambda_{\min} < \lambda_0$ must be orthogonal to the ground state $\psi_0(x)$, and this contradicts the positivity of both functions.
(2) The eigenvalues below λ_0 can exist if H does not have a ground state at λ_0. For instance, they definitely exist if one of the junctions is "wide enough" (in contrast to the O. Post condition[24]). The solution $u_\varepsilon(\tau,x)$ in this case has asymptotics different from the one stated in Theorem 6.1. In particular, if the function φ (see (31)) is positive, then

$$\varepsilon^2 \ln u_\varepsilon(\tau,x) \to \lambda_{\min}, \quad \tau \to \infty.$$

What is more important, the total mass of the heat energy in this case is concentrated in an arbitrarily small, as $\varepsilon \to 0$, neighborhood of the junctions. The limiting diffusion process on Γ degenerates.

Proof. For simplicity, we shall assume that the Dirichlet boundary condition is imposed on $\partial \Omega_\varepsilon$. Obviously, the function $\psi_0\left(\frac{x}{\varepsilon}\right)$ is the ground state in the spider domain Ω_ε. In particular,

$$\varepsilon^2 \Delta \psi_0 + \lambda_0 \psi_0 = 0; \; \psi_0\left(\frac{x}{\varepsilon}\right) = \varphi_0\left(\frac{y}{\varepsilon}\right)[\rho_j + o(1)], \; x \in C_{j,\varepsilon}, \; |x| \to +\infty;$$

$$\psi_0|_{\partial \Omega_\varepsilon} = 0.$$

Put $u_\varepsilon(\tau,x) = \psi_0\left(\frac{x}{\varepsilon}\right) e^{-\frac{\lambda_0 \tau}{\varepsilon^2}} w_\varepsilon(\tau,x)$. Then

$$\frac{\partial w_\varepsilon}{\partial \tau} = \Delta w_\varepsilon + \frac{2}{\varepsilon} \nabla \left(\ln \psi_0\left(\frac{x}{\varepsilon}\right)\right) \cdot \nabla w_\varepsilon,$$

$$w_\varepsilon(0,x) = \varphi(\gamma)\theta\left(\frac{x}{\varepsilon}\right), \; \theta\left(\frac{x}{\varepsilon}\right) = \frac{1}{\rho_j} + o(1), \; x \in C_j, \; |x| \to +\infty. \quad (33)$$

We look for bounded solutions u_ε, w_ε of the parabolic problems. We do not need to impose boundary conditions on $\partial \Omega_\varepsilon$ on the function w_ε since the boundedness of w_ε implies that $u_\varepsilon = 0$ on $\partial \Omega_\varepsilon$. The parabolic problem (33) has a unique bounded solution (without boundary conditions on $\partial\Omega_\varepsilon$) since $\nabla_z \ln \psi_0 (\cdot)$ grows near $\partial\Omega_\varepsilon$. The growth of the coefficient in (33) does not allow the heat energy (or diffusion) to reach $\partial\Omega_\varepsilon$. The fundamental solution $q_\varepsilon = q_\varepsilon(\tau, x_0, x)$ of the problem (33) exists, is unique, and $\int_{\Omega_\varepsilon} q_\varepsilon(\tau, x_0, x)\, dx = 1$. This fundamental solution is the transition density of the Markov diffusion process $X_\tau^{(\varepsilon)} = (T_\tau^{(\varepsilon)}, Y_\tau^{(\varepsilon)})$ in Ω_ε with the generator $\tilde{\mathcal{H}}_\varepsilon = \Delta + \dfrac{2}{\varepsilon}\left(\nabla \ln \psi_0 \left(\dfrac{x}{\varepsilon}\right), \nabla\right)$.

Let $\tilde{\mathcal{H}} = \tilde{\mathcal{H}}_\varepsilon$, and $q = q_\varepsilon$ when $\varepsilon = 1$. The coefficients of the operator $\tilde{\mathcal{H}}$ are singular at the boundary of the domain. However, the transition density $q(\tau, x_0, x)$ does not vanish inside Ω. To be more exact, the Döblin condition holds, i.e. for any compact $\omega' \subset \omega$, there exist $\delta > 0$ such that for any channel C_j the following estimate holds

$$q(\tau, x_0, x) > \delta \quad \text{when} \quad T \geq \tau \geq 1,\ x = (t, y),\ x_0 = (t, y_0),\ y, y_0 \in \omega'.$$

The operator $\tilde{\mathcal{H}} = \Delta + 2(\nabla \ln \psi_0(x), \nabla)$ has a unique (up to normalization) invariant measure. This measure has the density $\pi(x) = \psi_0^2(x)$. In fact, if $\tilde{\mathcal{H}} = \Delta + (\nabla A, \nabla)$ then $\tilde{\mathcal{H}}^* = \nabla - (\nabla A, \nabla) - (\Delta A)$, and one can easily check that $\tilde{\mathcal{H}}^* e^{A(x)} = 0$. If we put now $A(x) = 2\ln\psi_0(x)$, we get $\tilde{\mathcal{H}}^* \pi(x) = 0$.

When $t > \delta_0 > 0$ and $\varepsilon \to 0$ the transversal component $Y_\tau^{(\varepsilon)}$ and the longitudinal component $T_\tau^{(\varepsilon)}$ of the diffusion process in Ω_ε are asymptotically independent. The transversal component $Y_\tau^{(\varepsilon)}$ oscillates very fast and has asymptotically ($\varepsilon \to 0$) invariant measure $\dfrac{1}{\varepsilon}\varphi_0^2\left(\dfrac{x}{\varepsilon}\right)$. The latter follows from the Döblin condition. The longitudinal component has a constant diffusion with the drift which is exponentially small (of order $O(e^{-\frac{\gamma}{\varepsilon^2}})$, $\gamma > 0$) outside any neighborhood of the junction.

Under conditions above, one can apply (with minimal modifications) the fundamental averaging procedure by Freidlin-Wentzel[12] which leads to the convergence (in law on each compact interval in τ) of the distribution of the process $X_\tau^{(\varepsilon)}$ to the distribution of the process on Γ with the generator $\dfrac{d^2}{dt^2}$ on the space of functions on Γ smooth outside of the vertex and satisfying the appropriate GC. The GC are defined by the limiting invariant measure. This limiting measure on Γ is equal (up to a normalization) to ρ_j on edges Γ_j. This leads to the GC (32) of the generalized Kirchhoff form. The proof is complete. □

Theorem 6.2. *Let the operator H in a spider domain Ω with the Dirichlet or Robin condition at $\partial\Omega$ have a ground state at $\lambda = \lambda_0$, and let $\lambda =$*

$\lambda_0 + O(\varepsilon^2)$. Then the GC (17) has the generalized Kirchhoff form: ζ is continuous at the vertex v and $\sum_{j=1}^{d} \rho_j \frac{d\zeta}{dt_j}(0) = 0$.

This statement follows immediately from Theorem 6.1 since it is already established that the GC has the Dirichlet/Neumann form.

7. Effective potential

As it was already mentioned earlier, the GC (17) is λ-dependent. The following result allows one to reduce the original problem in Ω_ε to a Schrödinger equation on the limiting graph with arbitrary λ-independent GC and a λ-independent matrix potential. The potential depends on the choice of the GC. Only the lower part of the a.c. spectrum $\lambda_0 \leq \lambda \leq \lambda_1$ will be considered. It is assumed below that $\varepsilon = 1$, and the index ε is omitted everywhere.

Let $T(\lambda), \lambda \in [\lambda_0, \lambda_1]$, be the scattering matrix for a spider domain Ω, and let $\lambda_{-N} \leq \lambda_{-N+1} \leq \cdots \leq \lambda_{-1} < \lambda_0$ be the eigenvalues of the discrete spectrum of H below the threshold λ_0. The function $T(\lambda)$ has an analytic extension into the complex plane with the cut along $[0, \infty)$. It has poles at $\lambda = \lambda_{-N}, ..., \lambda_{-1}$. Let $m_{-N}, \cdots, m_{-1}$ be the corresponding residues (Hermitian $d \times d$ matrices). These residues contain complete information on the multiplicity of λ_j, $j = -N, \cdots -1$ and on the exponential asymptotics of the eigenfunctions $\psi_j(x)$, $|x| \to +\infty$.

Theorem 3.3 allows one to reduce the problem in Ω_ε to an equation for a function ζ on the limiting graph Γ with appropriate GC at the vertex. Consider the vector $\psi := \zeta^{(v)} = \zeta^{(v)}(t)$, $t \geq 0$, whose components are the restrictions of ζ to the edges of Γ. Note that the GC were formulated through the vector $\zeta^{(v)}$. Now we would also like to treat the equation for the function ζ on Γ as a linear system for the vector $\zeta^{(v)}$ on the half axis $t \geq 0$.

Theorem 7.1. *There exists an effective fast decreasing matrix $d \times d$ potential $V(t)$ such that $V(t) = V^*(t)$, and the problem*

$$-\psi'' + [V(t) - \lambda_0 I]\psi = \lambda\psi, \quad t \geq 0, \quad \psi(0) = 0 \qquad (34)$$

has the same spectral data on the interval $(-\infty, \lambda_1)$ as the original problem in Ω. The latter means that the scattering matrix $S(\lambda)$ of the problem (34) coincides with $T(\lambda)$ on the interval $[\lambda_0, \lambda_1]$, and the poles and residues of $S(\lambda)$ and $T(\lambda)$ are equal.

Remark 7.1.

(1) The potential is not uniquely defined.
(2) The Dirichlet condition $\psi(0) = 0$ can be replaced by any fixed GC, say the Kirchhoff one (of course, with the different effective potential).
(3) Different effective potentials appeared when explicitly solvable models were studied in our paper.[20]

Proof. This statement is a simple corollary of the inverse spectral theory by Agranovich and Marchenko for 1-D matrix Schrödinger operators.[1] One needs only to show that $T(\lambda)$ can be extended to the semiaxis $\lambda > \lambda_1$ in such a way that the extension will satisfy all the conditions required by the Agranovich-Marchenko theory. □

Example 7.1. For the statements of Lemma 4.1 and Theorem 6.2. Consider the Schrödinger operator $H = -\frac{d^2}{dt^2} + v(t)$ on the whole axis with a potential $v(t)$ compactly supported on $[-1, 1]$. This operator may serve as a simplified version of the operator (34). The simplest explicitly solvable model from Ref. 20 also leads to the operator H. The GC at $t = 0$ for this explicitly solvable model are determined by the limit, as $\varepsilon \to 0$, of the solution of the equation $H_\varepsilon \psi_\varepsilon = f$, where $H_\varepsilon = -\frac{d^2}{dt^2} + \varepsilon^{-2} v\left(\frac{t}{\varepsilon}\right)$ and f is compactly supported and vanishing in a neighborhood of $t = 0$. The solution ψ_ε is understood as L^2_{loc} limit of $(H_\varepsilon + i\mu)^{-1} f \in L^2$ as $\mu \to +0$.

Of course, $\lambda = 0$ is the bottom of the a.c. spectrum for H. If operator H does not have negative eigenvalues, then the equation $H\psi = 0$ has a unique (up to a constant factor) positive solution $\psi_0(t)$, which is not necessarily bounded. If this solution is linear outside $[-1, 1]$, then the limiting GC are the Dirichlet ones. This case is generic. If this solution is constant on one of the semiaxis, then the GC are the Dirichlet/Neumann conditions. Finally, if $\psi_0(t) = \rho_\pm$ for $\pm t \geq 1$, then we have the situation of Theorem 6.2: the ground state and the generalized Kirchhoff's GC.

One can get a nontrivial Kirchhoff's condition even in the case when H has a negative spectrum. It is sufficient to assume that $\lambda = 0$ is the eigenvalue (but not the minimal one) of the Neumann spectral problem for H on $[-1, 1]$.

References

1. Z. S. Agranovich, V. A. Marchenko. *The inverse problem of scattering theory*, Gordon and Breach Publishers, New-York, 1963.

2. G. Dell'Antonio, L. Tenuta. Quantum graphs as holonomic constraints. *J. Math. Phys.* **47** (2006), 072102:1–21.
3. P. Duclos, P. Exner. Curvature-induced bound states in quantum waveguides in two and three dimensions. *Rev. Math. Phys.* **7** (1995), 73–102.
4. P. Duclos, P. Exner, P. Stovicek. Curvature-induced resonances in a two-dimensional Dirichlet tube. *Ann. Inst. H. Poincaré* **62** (1995), 81–101
5. P. Exner, O. Post. Convergence of spectra of graph-like thin manifolds. *J. Geom. Phys.* **54** (2005), 77–115.
6. P. Exner , P. Šeba. Electrons in semiconductor microstructures: a challenge to operator theorists. *Schrödinger Operators, Standard and Nonstandard* (Dubna 1988), World Scientific, Singapure (1989), 79–100.
7. P. Exner, P. Šeba. Bound states in curved quantum waveguides. *J. Math. Phys.* **30** (1989), 2574–2580.
8. P. Exner, P. Šeba. Trapping modes in a curved electromagnetic waveguide with perfectly conducting walls. *Phys. Lett. A* **144** (1990), 347–350.
9. P. Exner, S. A. Vugalter. Asymptotic estimates for bound states in quantum waveguides coupled laterally through a narrow window. *Ann. Inst. H. Poincaré, Phys. Theor.* **65** (1996), 109–123.
10. P. Exner, S. A. Vugalter. On the number of particles that a curved quantum waveguide can bind. *J. Math. Phys.* **40** (1999), 4630–4638.
11. P. Exner, T. Weidl. Lieb-Thirring inequalities on trapped modes in quantum wires. *Proceedings of the XIII International Congress on Mathematical Physics* (London 2000); to appear [mp_arc 00-336].
12. M. Freidlin, A. Wentzel. Diffusion processes on graphs and averaging principle. *Ann. Probab.* **21 (4)** (1993), 2215–2245.
13. M. Freidlin. *Markov Processes and Differential Equations: Asymptotic Problems.* Lectures in Mathematics, ETH Zurich, Birkhäuser Verlag, Basel, 1996.
14. V. Kostrykin, R. Schrader. Kirchhoff's rule for quantum waves. *J. Phys. A: Mathematical and General* **32** (1999), 595–630.
15. P. Kuchment. Graph models of wave propagation in thin structures. *Waves in Random Media* **12** (2002), 1–24.
16. P. Kuchment. Quantum graphs. I. Some basic structures. *Waves in Random Media* **14 (1)** (2004), 107–128.
17. P. Kuchment. Quantum graphs. II. Some spectral properties of quantum and combinatorial graphs. *Journal of Physics A: Mathematical and General* **38 (22)** (2005), 4887–4900.
18. P. Kuchment, H. Zeng. Convergence of spectra of mesoscopic systems collapsing onto a graph. *J. Math. Anal. Appl.* **258** (2001), 671–700.
19. P. Kuchment, H. Zeng. Asymptotics of spectra of Neumann Laplacians in thin domains. *Advances in Differential Equations and Mathematical Physics*, Yu. Karpeshina etc (Editors), Contemporary Mathematics, AMS, 387 (2003), 199–213.
20. S. Molchanov, B. Vainberg. Transition from a network of thin fibers to quantum graph: an explicitly solvable model. *Contemporary Mathematics, AMS* **415** (2006), 227–240.

21. S. Molchanov, B. Vainberg. Scattering solutions in networks of thin fibers: small diameter asymptotics. *Comm. Math. Phys.* **273 (2)** (2007), 533–559.
22. A. Mikhailova, B. Pavlov, I. Popov, T. Rudakova, A. Yafyasov. Scattering on a compact domain with few semi-infinite wires attached: resonance case. *Math. Nachr.* **235** (2002), 101–128.
23. B. Pavlov, K. Robert. Resonance optical switch: calculation of resonance eigenvalues. *Waves in periodic and random media* (South Hadley, MA, 2002), Contemporary Mathematics, AMS, 339 (2003), 141–169.
24. O. Post. Branched quantum wave guides with Dirichlet BC: the decoupling case. *Journal of Physics A: Mathematical and General* **38 (22)** (2005), 4917–4932.
25. O. Post. Spectral convergence of non-compact quasi-one-dimensional spaces. *Ann. Henri Poincaré* **7** (2006), 933–973.
26. J. Rubinstein, M. Schatzman. Variational problems on multiply connected thin strips. I. Basic estimates and convergence of the Laplacian spectrum. *Arch. Ration. Mech. Anal.* **160 (4)** (2001), 293–306.

Adiabatic limits and quantum decoherence

Rolando Rebolledo*

Laboratorio de Análisis Estocástico
Facultad de Matemáticas
Pontificia Universidad Católica de Chile
Casilla 306, Santiago 22, Chile
E-mail: rrebolle@uc.cl

Dominique Spehner

Institut Fourier
Université de Grenoble
100 rue des Maths
BP 74, 38402 Saint Martin d'Hères, France

Quantum Markov semigroups have been used as a predominant model of open quantum dynamics where the system interacts with the environment. Thus, the main dynamics is submitted to perturbations which change the nature of its mathematical structure. This paper explores a class of perturbations which do not change the invariant elements of the main dynamics, which we call *adiabatic perturbations*. This is typically the case when the environment evolves according to a slower time scale than that of the system dynamics. As an illustration, we study a class of interacting limit behaviour leading to adiabatic perturbations of the main dynamics. Finally, we study the connection between adiabatic perturbations and the phenomenon of *quantum decoherence*.

Keywords: Quantum Markov semigroups, classical reductions, adiabatic perturbations, quantum exclusion semigroup

1. Introduction

Consider the space $\mathfrak{h} = \mathbb{C}^2$ and the basis $e_0 = \begin{pmatrix} 1 \\ 0 \end{pmatrix}$, $e_1 = \begin{pmatrix} 0 \\ 1 \end{pmatrix}$. Call

$$H = |e_1\rangle\langle e_1| = \begin{pmatrix} 0 & 0 \\ 0 & 1 \end{pmatrix}.$$

---

*Work partially supported by FONDECYT grant 1030552 and grant ACT 13 of the Bicentennial Foundation

A (closed) quantum dynamics associated to the Hamiltonian H is defined through a group of unitary operators $U_t : \mathfrak{h} \to \mathfrak{h}$ given by $U_t = \exp(-itH)$, $(t \in \mathbb{R})$, that is,

$$U_t = \begin{pmatrix} 1 & 0 \\ 0 & e^{-it} \end{pmatrix}.$$

Equivalently, the operator H defines an automorphism group α_t on the algebra $\mathfrak{L}(\mathfrak{h})$ of all linear (bounded) operators on $\mathfrak{h}$ which is isomorphic with $M_2(\mathbb{C})$ the algebra of two by two complex matrices:

$$\alpha_t(x) = U_t^* x U_t, \ (t \in \mathbb{R}).$$

This is the so called *Heisenberg picture* of the dynamics, while its dual version $\alpha_{*t}(\rho) = U_t \rho U_t^*$ defined on unit trace operators ρ, bears the name of Schrödinger.

Consider a positive operator with unit trace

$$\rho = \begin{pmatrix} p & re^{i\theta} \\ re^{-i\theta} & q \end{pmatrix},$$

where $r > 0$, $\theta \in [0, 2\pi]$, $0 \leq r \leq 1/2$ and $(p-q)^2 \leq 1 - 4r^2$, $p, q > 0$, $p + q = 1$.

The evolution of ρ at time t is then given by

$$\alpha_{*t}(\rho) = \begin{pmatrix} p & re^{i(t+\theta)} \\ re^{-i(t+\theta)} & q \end{pmatrix}.$$

Notice that the off-diagonal terms do not disappear as $t \to \infty$. These terms are known as the *coherences* in Quantum Mechanics, since they carry the non-commutative characteristics of the physical model.

The generator of the group α_* (respectively α) is $\delta_*(\rho) = i[\rho, H]$, (resp. $\delta(x) = i[H, x]$ for all endomorphism x).

And the set of fixed points of α is

$$\mathcal{F}(\alpha) = \{x : [H, x] = 0\}$$
$$= \left\{ x : x = f(|e_1\rangle\langle e_1|) = \begin{pmatrix} f(0) & 0 \\ 0 & f(1) \end{pmatrix}, \text{ for some function } f \text{ on } \{0, 1\} \right\}.$$

Now we will open the system, that is, interaction with the environment is allowed, so that it is embedded as a "small subsystem" in a bigger structure that we call the *total system*. Within this structure, the environment is supposed to be a "big" subsystem which is in equilibrium, represented by an equilibrium state ρ_β. The interaction of our small subsystem with the environment introduces a perturbation in both of them. We will assume that the environment returns to equilibrium much faster than the time scale of the small system evolution.

Given a state ρ on the initial space $\mathfrak{h}$, the reduced dynamics is obtained by performing a limit procedure on the time scale evolution of the environment and a partial trace of the total dynamics. This is the so called *Markov approximation* of the open system dynamics. The reader is referred to the book of Accardi, Lu and Volovich (see Ref. 1) for further details and historical references on this matter. To give a rough picture of the approximation, under suitable hypotheses one obtains a limit dynamics $\widetilde{U}_t$ defined on the space $\mathfrak{h} \otimes \mathfrak{h}_R$, where $\mathfrak{h}_R$ is the Hilbert space associated to the reservoir. Then the evolution of the state ρ is given by the following partial trace on the reservoir variables:

$$\mathcal{T}_{*t}(\rho) = \operatorname{tr}_R \left(\widetilde{U}_t \rho \otimes \rho_\beta \widetilde{U}_t^* \right). \tag{1}$$

The dual version of the above expression gives the evolution $\mathcal{T}_t(x)$ of any observable x:

$$\operatorname{tr}(\mathcal{T}_{*t}(\rho)x) = \operatorname{tr}(\rho \mathcal{T}_t(x)), \ (t \geq 0).$$

It turns out that $\mathcal{T}$ above defines a semigroup structure on $M_2(\mathbb{C})$. This semigroup (resp. its dual $\mathcal{T}_*$) has a generator that can be written in the form

$$\mathcal{L}(x) = i[H, x] + \mathcal{D}(x), \tag{2}$$

(resp. $\mathcal{L}_*(\rho) = i[\rho, H] + \mathcal{D}_*(\rho)$), where $\mathcal{D}$ (resp. $\mathcal{D}_*$) represents the dissipation due to the interaction of the system with the reservoir.

So that, for instance, assume that the dissipation is written in the so called *Lindblad form* (cf. Ref. 7) as follows:

$$\mathcal{D}(x) = -\frac{1}{2} \left(\sigma_+ \sigma_- x - 2\sigma_+ x \sigma_- + x \sigma_+ \sigma_- \right), \tag{3}$$

where

$$\sigma_+ = \begin{pmatrix} 0 & 0 \\ 1 & 0 \end{pmatrix},$$

$$\sigma_- = \begin{pmatrix} 0 & 1 \\ 0 & 0 \end{pmatrix}.$$

Notice that $\sigma_+ \sigma_- = H = |e_1\rangle\langle e_1|$, so that

$$\sigma_+ \sigma_- f(H) = f(H) \sigma_+ \sigma_- = H f(H).$$

Moreover, a simple computation shows that

$$\sigma_+ f(H) \sigma_- = f(0) H.$$

And since $[H, f(H)] = 0$, we obtain that

$$\mathcal{L}(f(H)) = f(0)H - Hf(H) = g(H),$$

where $g(y) = f(0)y - yf(y)$, for all $y \in \{0, 1\}$. So that $\mathcal{L}(f(H)) \in \mathcal{F}(\alpha)$ and the perturbation introduced by the environment in the main dynamics leaves its set of invariant elements unchanged.

We say that the environment induces an *adiabatic perturbation* on the system. Physical examples of this kind of perturbations have been retrieved in all quantum dynamical systems obtained via a weak coupling (or van Hove) limit (see Refs. 4,14) as well as in the particular case of quantum jumps models studied by Spehner and Bellissard (see Refs. 11,12) and those obtained via the so-called stochastic limit by Accardi, Lu and Volovich (see Ref. 1). An important consequence of this type of perturbation is the classical reduction of the dynamics by the commutative algebra $\mathcal{F}(\alpha)$, that is, since this algebra remains invariant under $\mathcal{T}$, one obtains a classical Markov semigroup via the Spectral Theorem due to von Neumann. Indeed, since the algebra $\mathcal{F}(\alpha)$ coincides with the commutant of H, in the current example this commutant coincides with the algebra generated by H, so that it is isomorphic to the algebra of bounded functions defined on $\{0, 1\}$. That means that given any bounded function f defined on $\{0, 1\}$, the semigroup $\mathcal{T}_t(\cdot)$ associates to the operator $f(H)$ an operator $g(H)$ where g is another bounded function on $\{0, 1\}$. So that, we can define $\mathcal{T}_t f := g$ and we have the representation

$$\mathcal{T}_t f(H) = \mathcal{T}_t(f(H)).$$

Denote $\mathfrak{A}$ the commutative algebra of all bounded functions on $\{0, 1\}$, then $\mathcal{T}_t : \mathfrak{A} \to \mathfrak{A}$, for all $t \geq 0$, defines a classical Markov semigroup.

In addition, it is clear that the operator $\rho_\infty = |e_0\rangle\langle e_0|$, the projection over the space generated by e_0, is $\mathcal{T}_*$-invariant, and one can prove that (see Ref. 5) for any other state ρ, it holds

$$\text{tr}(\rho \mathcal{T}_t(x)) = \text{tr}(\mathcal{T}_{*t}(\rho)x) \to \text{tr}(\rho_\infty x), \qquad (4)$$

for all $x \in M_2(\mathbb{C})$. In particular, one obtains

$$\langle e_0, \mathcal{T}_{*t}(\rho)e_1 \rangle \to 0$$
$$\langle e_1, \mathcal{T}_{*t}(\rho)e_0 \rangle \to 0.$$

The above equations are interpreted in Physics as a loss of *coherence*, that is, any quantum state evolves towards a classical (commutative) state.

The goal of the current paper is to study a notion of adiabatic perturbation of a quantum dynamics, as well as its relationship with the appearance of a classical Markov structure in open quantum systems.

2. Adiabatic perturbations

Consider a complex separable Hilbert space $\mathfrak{h}$ and denote $\mathfrak{L}(\mathfrak{h})$ the algebra of all its bounded operators. We denote $\alpha = (\alpha_t)_{t \in \mathbb{R}_+}$ a one parameter group of *-automorphisms of $\mathfrak{L}(\mathfrak{h})$. We start by the simplest situation in which α is norm-continuous. Then its generator δ is an everywhere defined symmetric derivation of $\mathfrak{L}(\mathfrak{h})$, so that there exists an $H = H^* \in \mathfrak{L}(\mathfrak{h})$ such that

$$\alpha_t(x) = e^{itH} x e^{-itH}, \tag{5}$$

for all $x \in \mathfrak{L}(\mathfrak{h})$ and $t \in \mathbb{R}$. The algebra generated by the operator H is denoted $W^*(H)$ and its generalised commutator by $W^*(H)'$. The latter coincides with the former if and only if H is non-degenerate or multiplicity-free (see Ref. 8, chap. 4, sect. 4.7); in that case $W^*(H)$ is a maximal commutative algebra.

Denote $\mathcal{F}(\alpha) = \{x \in \mathfrak{A} : \alpha_t(x) = x, \text{for all } t \in \mathbb{R}\}$ the set of *fixed points* or *invariants elements* of α. It holds that $\mathcal{F}(\alpha) = W^*(H)'$.

We recall the definition of a Quantum Markov Semigroup on $\mathfrak{L}(\mathfrak{h})$ here below. Our instrinsic probabilistic setting is non commutative. However, no independence concept is involved, and there is no need to go into the Voiculescu free probability framework.

Definition 2.1. A *Quantum Markov Semigroup* (QMS) $\mathcal{T} = (\mathcal{T}_t)_{t \in \mathbb{R}_+}$ on $\mathfrak{L}(\mathfrak{h})$ is a w^*-continuous semigroup of completely positive maps which preserves the unit.

The characterization of the generator $\mathcal{L}(\cdot)$ of a norm-continuous QMS $\mathcal{T}$ has been obtained by Lindblad (see Ref. 7), for semigroups defined on $\mathfrak{L}(\mathfrak{h})$ and by Christensen and Evans (see Ref. 3) in the C^*-case, that is, for norm-continuous semigroups defined on a general C^*-algebra $\mathfrak{A}$. We recall these results in our particular framework here below. Norm-continuity is a very strong condition which is not always satisfied by semigroups appearing in the applications to Physics. A condition which is less restrictive is that of *pointwise continuity in norm*, that is $\lim_{t \to 0} \|\mathcal{T}_t(x) - x\| = 0$ for all x in the algebra where the semigroup is defined. A *Quantum Feller Semigroup* is a QMS defined on a C^*-algebra which is pointwise continuous in norm as before. This is the extension of the classical Feller property to non-commutative Probability Theory.

Theorem 2.1. *If the Quantum Markov Semigroup $\mathcal{T}$ is norm-continuous on $\mathfrak{L}(\mathfrak{h})$, there exists a completely positive map $\Phi : \mathfrak{L}(\mathfrak{h}) \to \mathfrak{L}(\mathfrak{h})$ and a bounded operator G which is the generator of a contraction semigroup on $\mathfrak{h}$, such that the generator $\mathcal{L}$ of $\mathcal{T}$ is represented as*

$$\mathcal{L}(x) = G^*x + \Phi(x) + xG, \qquad (6)$$

for all $x \in \mathfrak{L}(\mathfrak{h})$, where $\Phi(1) = -2\Re(G)$. We refer to this formula as the Christensen-Evans form of the generator.

Moreover, using Kraus representation of completely positive maps, there exists bounded operators $H = H^*$, L_k, $(k \in \mathbb{N})$, such that $\sum_k L_k{}^* L_k \in \mathfrak{L}(\mathfrak{h})$ and the generator $\mathcal{L}$ is represented in Lindblad form as

$$\mathcal{L}(a) = i[H, x] - \frac{1}{2} \sum_k \left(L_k{}^* L_k x - 2 L_k{}^* x L_k + x L_k{}^* L_k \right), \qquad (7)$$

for all $x \in \mathfrak{L}(\mathfrak{h})$.

The two formulas are related by $G = -iH - \frac{1}{2} \sum_k L_k{}^* L_k$, $\Phi(x) = \sum_k L_k{}^* x L_k$, though it is well known that none of the above expressions is uniquely determined.

We will speak of H as the *bare* Hamiltonian, since it leads to the generator δ of the automorphism group α which represents the evolution of the main system. The dissipative term $\mathcal{D}(x) = -\frac{1}{2} \sum_k \left(L_k{}^* L_k x - 2 L_k{}^* x L_k + x L_k{}^* L_k \right)$ is due to the interaction between the system and the reservoir. In which follows we will consider that a bare Hamiltonian is given at the outset. That is, we know the isolated evolution of the main system. So that, the dissipation will be characterised as a perturbation of the given main system.

Here below we introduce the definition of an adiabatic perturbation of a given automorphism group, which means that the dissipative terms do not dramatically change the set of fixed points of the automorphism group which represents the bare system.

Definition 2.2. The quantum Markov semigroup is *an adiabatic perturbation* of an automorphism group α if it satisfies

$$\mathcal{T}_t(\mathcal{F}(\alpha)) \subseteq \mathcal{F}(\alpha), \qquad (8)$$

for all $t \geq 0$.

Theorem 2.2. *Assume that the quantum Markov semigroup $\mathcal{T}$ is norm-continuous on $\mathfrak{L}(\mathfrak{h})$ and denote $\mathcal{L}(\cdot)$ its generator represented in the Christensen-Evans form (6). Call $H = -\Im(G)$ and denote α the group of automorphisms with generator $\delta(x) = i[H, x]$.*

(1) If for all $x \in \mathcal{F}(\alpha)$ one has $\Phi(x) \in \mathcal{F}(\alpha)$, then $\mathcal{T}$ is an adiabatic perturbation of α.

(2) Reciprocally, assume that $\Phi(\mathbf{1}) \in \mathcal{F}(\alpha)$ and that the semigroup is an adiabatic perturbation of α, then $\Phi(x) \in \mathcal{F}(\alpha)$ for all $x \in \mathcal{F}(\alpha)$.

Proof. Since the semigroup is Markovian, it holds that $\mathcal{L}(\mathbf{1}) = G^* + \Phi(\mathbf{1}) + G = 0$, so that $\Re(G) = -\frac{1}{2}\Phi(\mathbf{1})$.

(1) Given any $x \in \mathcal{F}(\alpha)$, x commutes with $H = -\Im(G)$. Moreover, by hypothesis, $\Phi(x)$ commutes with H, in particular $[H, \Re(G)] = 0$, from which we obtain that $[H, G] = 0 = [H, G^*]$ and $\mathcal{L}(x)$ commutes with H as well. Thus $\mathcal{L}(x) \in \mathcal{F}(\alpha)$ if $x \in \mathcal{F}(\alpha)$ and the semigroup is an adiabatic perturbation of α.

(2) If $\Phi(\mathbf{1})$ commutes with H, it follows that $[H, G] = 0$. Now, given any $x \in \mathcal{F}(\alpha)$, $\mathcal{L}(x) - G^*x - xG = \Phi(x)$ is also an element of $\mathcal{F}(\alpha)$ since the semigroup is an adiabatic perturbation of α. □

Corollary 2.1. *Assume that the semigroup $\mathcal{T}$ is norm-continuous. Suppose in addition that all the coefficients L_k of (7) satisfy the following hypothesis:*

(H) There exists a matrix $(c_{k,\ell})$ of bounded maps from $W^*(H)' \to W^*(H)'$ such that $c_{k,\ell}(a)^* = c_{\ell,k}(a^*)$, for all $a \in W^*(H)'$ and such that

$$[H, L_k] = \sum_{\ell \in \mathbb{N}} c_{k,\ell}(H) L_\ell, \quad (k \in \mathbb{N}). \tag{9}$$

Then $\mathcal{T}$ is an adiabatic perturbation of α.

Proof. The algebra $\mathcal{F}(\alpha)$ coincides with the generalized commutator algebra $W^*(H)'$ of the von Neumann algebra $W^*(H)$ generated by H. Notice that (9) implies that

$$[H, L_k^* x L_k] = 0,$$

for all k if x commutes with H. Indeed, since $[H, \cdot]$ is a derivation, it turns out that

$$[H, L_k^* x L_k] = -\sum_\ell L_\ell^* c_{\ell,k}(H) x L_k + \sum_\ell L_k^* x c_{k,\ell}(H) L_\ell.$$

Now, exchanging ℓ by k in the first sum and noticing that x and $c_{k,\ell}(H)$ commute yields the result.

Therefore, $\Phi(x) = \sum_k L_k^* x L_k$ satisfies the hypothesis (1) of the previous theorem and the proof is completed. □

Remark 2.1. Recently Fagnola and Skeide have posted a preprint (see Ref. 6) proving that (H) is also a necessary condition to leave the algebra $W^*(H)'$ invariant under the action of the semigroup $\mathcal{T}$. Their method of proof is based on Hilbert modules.

3. Adiabatic limits

Now, assume that there exists a faithful normal stationary state for the automorphism group α. Then $\mathcal{F}(\alpha)$ is a von Neumann algebra, and there exists a conditional expectation $\mathbb{E}^{\mathcal{F}(\alpha)}(\cdot)$. Indeed, $\mathcal{F}(\alpha)$ is invariant under the action of the modular group defined in the theory of Tomita and Takesaki: $x \mapsto \sigma_t(x) = \rho^{it} x \rho^{-it}$, where ρ is the density matrix associated with the faithful normal stationary state, $x \in \mathfrak{L}(\mathfrak{h})$, $t \in \mathbb{R}$. Thus, as a result of the ergodic theory of the modular group the conditional expectation $\mathbb{E}^{\mathcal{F}(\alpha)}(\cdot)$ with respect to $\mathcal{F}(\alpha)$ exists and it is an orthogonal projection for the scalar product associated to the invariant state: $(x, y) \mapsto \operatorname{tr}(\rho x^* y)$, $x, y \in \mathfrak{L}(\mathfrak{h})$. In this case, we obtain the following property.

Proposition 3.1. *Under the assumption before, given any bounded completely positive linear map $\mathcal{T}$ on $\mathfrak{L}(\mathfrak{h})$ such that $\|\mathcal{T}(x)\| \leq \|x\|$, the family*

$$\mathcal{C}(x) = \left(\frac{1}{T}\int_0^T \alpha_t \circ \mathcal{T} \circ \alpha_{-t}(x) dt; \ T \geq 0\right),$$

admits a w^*-limit on $\mathfrak{L}(\mathfrak{h})$, denoted $\mathcal{T}^\alpha(x)$, for all $x \in \mathfrak{L}(\mathfrak{h})$. The map $x \mapsto \mathcal{T}^\alpha(x)$ is completely positive and bounded.

Proof. We first use Banach-Alaglou Theorem. Given $x \in \mathfrak{L}(\mathfrak{h})$,

$$\left\|\frac{1}{T}\int_0^T \alpha_t \circ \mathcal{T} \circ \alpha_{-t}(x) dt\right\| \leq \|x\|.$$

So that, given any state ω,

$$\omega\left(\frac{1}{T}\int_{-T}^T \alpha_t \circ \mathcal{T} \circ \alpha_{-t}(x) dt\right) \leq \|x\|,$$

for all $T \geq 0$, so that $\mathcal{C}(x)$ is w^*-compact. Take two different w^*-limit points $T_1^\alpha(x)$ and $T_2^\alpha(x)$ of $\mathcal{C}(x)$. Since complete positivity is preserved by composition of maps and by w^*-limits, $x \mapsto T_j^\alpha(x)$ is completely positive for $j = 1, 2$.

If ω_0 is a faithful normal invariant state, it follows easily that

$$\omega_0(T_1^\alpha(x)) = \omega_0(T \circ \mathbb{E}^{\mathcal{F}(\alpha)}(x)) = \omega_0(T_2^\alpha(x)).$$

Since ω_0 is faithful, the above equation, together with the positivity of maps imply that $T_1^\alpha(x) = T_2^\alpha(x)$. □

This proposition allows to introduce a procedure for building up *adiabatic approximations* of a given semigroup.

Definition 3.1. Given any quantum Markov semigroup T on $\mathfrak{L}(\mathfrak{h})$, we define its *adiabatic limit* as the semigroup $(T^\alpha{}_t)_{t \in \mathbb{R}^+}$ where each T_t^α is obtained through Proposition 3.1

Proposition 3.2. *If there exists a faithful normal invariant state, the adiabatic limit of a QMS is an adiabatic perturbation of the automorphism group defined by its bare Hamiltonian.*

Proof. Let us keep the notations of Proposition 3.1. Then, for any $x \in \mathcal{F}(\alpha)$, $\alpha_t \circ \mathcal{T}_s(\alpha_{-t}(x)) = \alpha_t \circ \mathcal{T}_s(x)$, for all $s, t \geq 0$. Therefore, $T_s^\alpha(x) = \mathbb{E}^{\mathcal{F}(\alpha)}(\mathcal{T}_s(x)) \in \mathcal{F}(\alpha)$, for all $s \geq 0$. □

Adiabatic limits are used in numerous phenomenological descriptions of open quantum systems. As such, in view of the previous proposition, those approaches yield to quantum Markov semigroups endowed with a property of classical reduction by the algebra of fixed points of the main dynamics, the automorphism group generated by its bare Hamiltonian.

4. Invariant states and adiabatic perturbations. Decoherence

Theorem 4.1. *Assume that a QMS T is an adiabatic perturbation of an automorphism group generated by a non-degenerate bounded self-adjoint operator H. Then the restriction of T to the algebra $\mathcal{F}(\alpha)$ is isomorphic to a classical Markov semigroup $T = (T_t)_{t \in \mathbb{R}^+}$ defined on the algebra $L^\infty(\sigma(H), \nu)$ where $\sigma(H)$ is the spectrum of H and ν is the measure determined by its spectral decomposition. Moreover, if the semigroup T is norm-continuous, then T satisfies the Feller property, that is, the C^*-algebra*

$C_0(\sigma(H))$ of all continuous functions vanishing at infinity is invariant under T.

In particular, suppose that the semigroup $\mathcal{T}$ satisfies the hypothesis (H). Then a density matrix ρ which commutes with H defines an invariant state for $\mathcal{T}$ if and only if:

$$\mathrm{tr}(\rho\Phi(x)) = \mathrm{tr}(\rho\Phi(\mathbf{1})x), \tag{10}$$

for all $x \in \mathfrak{L}(\mathfrak{h})$, where Φ is related to the generator of $\mathcal{T}$ by (6).

Proof. If H is non degenerate, then the algebra $W^*(H)$ is maximal abelian (see Ref. 8 chap.4), therefore, $W^*(H) = W^*(H)' = \mathcal{F}(\alpha)$. Thus, by the Spectral Theorem, $\mathcal{F}(\alpha)$ is isomorphic to the space $L^\infty(\sigma(H), \mu)$ where μ is a Radon measure obtained from the spectral decomposition of H. More precisely, there exists an isometry $U : L^2(\sigma(H), \mu) \to \mathfrak{h}$ such that $f \mapsto UM_f U^*$ is an isometric *-ismorphism of $L^\infty(\sigma(H), \mu)$ on $\mathcal{F}(\alpha)$, where M_f denotes the operator multiplication by f. We define the semigroup T by

$$M_{T_t f} = \mathcal{T}_t(UM_f U^*).$$

This is a Markov semigroup, since the complete positivity is preserved, $\mathbf{1} \in \mathcal{F}(\alpha)$ and $T_t \mathbf{1} = 1$. If the original semigroup is norm continuous, then T is a contraction for the uniform norm, so that $C_0(\sigma(H))$ is invariant under the action of T and the semigroup is Feller.

Finally, notice that if ρ commutes with H the hypothesis (H) implies that it also commutes with $\Phi(\mathbf{1})$ which is an element of the abelian algebra $W^*(H)$. So that,

$$\mathrm{tr}(\mathcal{L}_*(\rho)x) = \mathrm{tr}(\rho\mathcal{L}(x))$$
$$= \mathrm{tr}(\rho G^*x) + \mathrm{tr}(\rho\Phi(x)) + \mathrm{tr}(\rho xG)$$
$$= \mathrm{tr}(\rho G^*x) + \mathrm{tr}(\rho\Phi(x)) + \mathrm{tr}(\rho Gx)$$
$$= \mathrm{tr}(\rho(G^*+G)x) + \mathrm{tr}(\rho\Phi(x))$$
$$= -\mathrm{tr}(\rho\Phi(\mathbf{1})x) + \mathrm{tr}(\rho\Phi(x)).$$

for all $x \in \mathfrak{L}(\mathfrak{h})$. Now, ρ is a fixed point for the predual semigroup $\mathcal{T}_*$ (invariant state) if and only if $\mathcal{L}_*(\rho) = 0$. The previous computation shows that the latter is equivalent to have (10) holds true. □

Corollary 4.1. *Under the hypothesis (H) and Lindblad notation of the generator of the quantum Markov semigroup of the previous Theorem, assume in addition that*

$$\{L_k, L_k^*\}' = \mathbb{C}\mathbf{1}.$$

Then the invariant state defined by a faithful density matrix $\rho = p(H)$ which satisfies (10) is unique and for all other density matrix σ it holds that $\mathcal{T}_{*t}(\sigma) \to \rho$ in the trace norm, as $t \to \infty$.

As a result, suppose that $(e_n)_{n \in \mathbb{N}}$ is an orthonormal basis of eigenvectors of H, then for all $n, m \in \mathbb{N}$, $n \neq m$, it holds that

$$\langle e_m, \mathcal{T}_{*t}(\sigma) e_n \rangle \to 0,$$

as $t \to \infty$. That is, the system is decoherent.

Proof. The hypothesis introduced here implies the ergodicity of the semigroup (see Ref. 5). Since ρ is supposed to satisfy (10), it is invariant. Moreover, since it is faithful and the system is ergodic, it holds that $\mathcal{T}_{*t}(\sigma) \to \rho$ in the trace norm, as $t \to \infty$. Finally, ρ is diagonal in the basis of eigenvectors of H, so that

$$\langle e_m, \mathcal{T}_{*t}(\sigma) e_n \rangle \to 0,$$

for all $n \neq m$. □

5. The quantum exclusion semigroup

The generator of this example is constructed via a second quantization procedure. For further details on this example the reader is referred to Ref. 9. Consider first a self-adjoint bounded operator H_0 defined on a separable complex Hilbert space $\mathfrak{h}_0$. H_0 will be thought of as describing the dynamics of a single fermionic particle. We assume that there is an orthonormal basis $(\psi_n)_{n \in \mathbb{N}}$ of eigenvectors of H_0, and denote E_n the eigenvalue of ψ_n ($n \in \mathbb{N}$). The set of all finite subsets of $\mathbb{N}$ is denoted $\mathfrak{P}_f(\mathbb{N})$ and for any $\Lambda \in \mathfrak{P}_f(\mathbb{N})$, we denote $\mathfrak{h}_0^\Lambda$ the finite-dimensional Hilbert subspace of $\mathfrak{h}_0$ generated by the vectors $(\psi_n;\ n \in \Lambda)$. To deal with a system of infinite particles we introduce the fermionic Fock space $\mathfrak{h} = \Gamma_f(\mathfrak{h}_0)$ associated to $\mathfrak{h}_0$ whose construction we recall briefly (see Ref. 2 for more detail).

The multi-particle Hilbert space associated to $\mathfrak{h}_0$ is the direct sum

$$\Gamma(\mathfrak{h}_0) = \bigoplus_{n \in \mathbb{N}} \mathfrak{h}_0^{\otimes n},$$

where $\mathfrak{h}_0^{\otimes n}$ is the n-fold tensor product of $\mathfrak{h}_0$, with the convention $\mathfrak{h}_0^{\otimes 0} = \mathbb{C}$. Define an operator $\mathbf{P}_a$ on $\Gamma(\mathfrak{h}_0)$) as follows,

$$\mathbf{P}_a(f_1 \otimes f_2 \otimes \ldots \otimes f_n) = \frac{1}{n!} \sum_\pi \varepsilon_\pi f_{\pi_1} \otimes \ldots \otimes f_{\pi_n}.$$

The sum is over all permutations $\pi : \{1,\ldots,n\} \to \{\pi_1,\ldots,\pi_n\}$ of the indices and ε_π is 1 if π is even and -1 if π is odd. Define the anti-symmetric tensor product on the Fock space as $f_1 \wedge \ldots \wedge f_n = \mathbf{P}_a(f_1 \otimes f_2 \otimes \ldots \otimes f_n)$. In this manner, the Fermi-Fock space $\mathfrak{h}$ is obtained as

$$\mathfrak{h} = \Gamma_f(\mathfrak{h}_0) = \mathbf{P}_a \left(\bigoplus_{n \in \mathbb{N}} \mathfrak{h}_0^{\otimes n} \right) = \bigoplus_{n \in \mathbb{N}} \mathfrak{h}_0^{\wedge n}.$$

We follow Ref. 2 to introduce the so-called fermionic *Creation* $b^\dagger(f)$ and *Annihilation* $b(f)$ operators on $\mathfrak{h}$, associated to a given element f of $\mathfrak{h}_0$. Firstly, on $\Gamma(\mathfrak{h}_0)$ we define $a(f)$ and $a^\dagger(f)$ by initially setting $a(f)\psi^{(0)} = 0$, $a^\dagger(f)\psi^{(0)} = f$, and for $\psi = (\psi^{(0)}, \psi^{(1)}, \ldots) \in \Gamma(\mathfrak{h}_0)$ with $\psi^{(j)} = 0$ for all $j \geq 1$,

$$a^\dagger(f)(f_1 \otimes \ldots \otimes f_n) = \sqrt{n+1} f \otimes f_1 \otimes \ldots \otimes f_n. \quad (11)$$

$$a(f)(f_1 \otimes \ldots \otimes f_n) = \sqrt{n} \langle f, f_1 \rangle f_2 \otimes f_3 \otimes \ldots \otimes f_n. \quad (12)$$

Finally, define annihilation and creation on $\Gamma_f(\mathfrak{h}_0)$ as $b(f) = \mathbf{P}_a a(f) \mathbf{P}_a$ and $b^\dagger(f) = \mathbf{P}_a a^\dagger(f) \mathbf{P}_a$. These operators satisfy the Canonical Anti-commutation Relations (CAR) on the Fermi-Fock space:

$$\{b(f), b(g)\} = 0 = \{b^\dagger(f), b^\dagger(g)\} \quad (13)$$

$$\{b(f), b^\dagger(g)\} = \langle f, g \rangle \mathbf{1}, \quad (14)$$

for all $f, g \in \mathfrak{h}_0$, where we use the notation $\{A, B\} = AB + BA$ for two operators A and B.

Moreover, $b(f)$ and $b^\dagger(g)$ have bounded extensions to the whole space $\mathfrak{h}$ since $\|b(f)\| = \|f\| = \|b^\dagger(f)\|$.

To simplify notations, we write $b_n^\dagger = b^\dagger(\psi_m)$ (respectively $b_n = b(\psi_n)$) the creation (respectively annihilation) operator associated with ψ_n in the space $\mathfrak{h}_0$, $(n \in \mathbb{N})$.

The C^*-algebra generated by $\mathbf{1}$ and all the $b(f)$, $f \in \mathfrak{h}_0$, is denoted $\mathfrak{A}(\mathfrak{h}_0)$ and it is known as the canonical *CAR* algebra.

Remark 5.1. *The algebra $\mathfrak{A}(\mathfrak{h}_0)$ is the unique, up to $*$-isomorphism, C^*-algebra generated by elements $b(f)$ satisfying the anti-commutation relations over $\mathfrak{h}_0$ (see e.g. Ref. 2, Theorem 5.2.5).*

Remark 5.2. *It is worth mentioning that the family $(b(f), b^\dagger(g); f, g \in \mathfrak{h}_0)$ is irreducible on $\mathfrak{h}$, that is, the only operators which commute with this family are the scalar multiples of the identity (see Ref. 2, Prop.5.2.2). Clearly, the same property is satisfied by the family $(b_n, b_n^\dagger; n \in \mathbb{N})$, since $(\psi_n)_{n \in \mathbb{N}}$ is an orthonormal basis of $\mathfrak{h}_0$.*

Remark 5.3. *The algebra* $\mathfrak{A}(\mathfrak{h}_0)$ *is the strong closure of* $\mathfrak{D} = \bigcup_{\Lambda \in \mathfrak{P}_f(\mathbb{N})} \mathfrak{A}(\mathfrak{h}_0^\Lambda)$ *(see Ref. 2, Proposition 5.2.6), this is the* quasi-local *property. Moreover, the finite dimensional algebras* $\mathfrak{A}(\mathfrak{h}_0^\Lambda)$ *are isomorphic to algebras of matrices with complex components.*

An element η of $\{0,1\}^\mathbb{N}$ will be called a *configuration* of particles. For each n, $\eta(n)$ will take the value 1 or 0 depending on whether the n-th site is occupied by a particle in the configuration η. In other terms, we say that the site n is occupied by the configuration η if $\eta(n) = 1$. We denote **S** the set of configurations η with a finite number of 1's, that is $\sum_n \eta(n) < \infty$. Each $\eta \in \mathbf{S}$ is then identifiable to the characteristic function $1_{\{s_1,\ldots,s_m\}}$ of a finite subset of $\mathbb{N}$, which, in addition, we will suppose ordered as $0 \leq s_1 < s_2 < \ldots < s_m$. For simplicity we write 1_k the configuration $1_{\{k\}}$, ($k \in \mathbb{N}$). Furthermore, we define

$$\mathbf{b}^\dagger(\eta) = b^\dagger_{s_m} b^\dagger_{s_{m-1}} \ldots b^\dagger_{s_1} \tag{15}$$

$$\mathbf{b}(\eta) = b_{s_m} b_{s_{m-1}} \ldots b_{s_1}, \tag{16}$$

for all $\eta = 1_{\{s_1,\ldots,s_m\}}$. Clearly, $\mathbf{b}^\dagger(1_k) = b^\dagger_k$, $\mathbf{b}(1_k) = b_k$, ($k \in \mathbb{N}$).

To obtain a cyclic representation of $\mathfrak{A}(\mathfrak{h}_0)$ we call $|0\rangle$ the vacuum vector in $\mathfrak{h}$, and $|\eta\rangle = \mathbf{b}^\dagger(\eta)|0\rangle$, ($\eta \in \mathbf{S}$). Then $(|\eta\rangle, \eta \in \mathbf{S})$ is an orthonormal basis of $\mathfrak{h}$. In this manner, any $x \in \mathfrak{A}(\mathfrak{h}_0)$ can be represented as an operator in $\mathfrak{L}(\mathfrak{h})$. Moreover, call $\mathfrak{v}$ the vector space spanned by $(|\eta\rangle, \eta \in \mathbf{S})$.

An elementary computation based on the C.A.R. shows that for any $\eta, \zeta \in \mathbf{S}$, it holds

$$b^\dagger_k|\eta\rangle = (1 - \eta(k))|\eta + 1_k\rangle, \tag{17}$$

$$b_k|\eta\rangle = \eta(k)|\eta - 1_k\rangle, \ (k \in \mathbb{N}). \tag{18}$$

We assume in addition that H_0 is **bounded from below**, so that there exists $b \in \mathbb{R}$ such that $b < E_n$ for all $n \in \mathbb{N}$. Then, the second quantization of H_0 becomes a self-adjoint operator H acting on $\mathfrak{h}$, with domain $D(H)$ which includes $\mathfrak{v}$ and can formally be written as

$$H = \sum_n E_n b^\dagger_n b_n. \tag{19}$$

It is worth mentioning that the restriction $H^\Lambda = \sum_{n \in \Lambda} E_n b^\dagger_n b_n$ of H to each space $\Gamma_f(\mathfrak{h}_0^\Lambda)$ is an element of the algebra $\mathfrak{A}(\mathfrak{h}_0^\Lambda)$, $\Lambda \in \mathfrak{P}_f(\mathbb{N})$, so that H^Λ is a bounded operator. Moreover, $\|H|\eta\rangle - H^\Lambda|\eta\rangle\| \to 0$ as Λ increases to $\mathbb{N}$, for each $\eta \in \mathbf{S}$.

The *transport* of a particle from a site i to a site j, at a rate $\gamma_{i,j} \geq 0$ with $\gamma_{i,i} = 0$, is described by an operator $L_{i,j}$ defined as

$$L_{i,j} = \sqrt{\gamma_{i,j}}\, b_j^\dagger b_i. \tag{20}$$

This corresponds to the action of a reservoir on the system of fermionic particles pushing them to jump between different sites. Each operator $L_{i,j}$ is an element of $\mathfrak{A}(\mathfrak{h}_0)$ and $\|L_{i,j}\| = \sqrt{\gamma_{i,j}}$. We additionnally assume that

$$\sup_i \sum_j \gamma_{i,j} < \infty. \tag{21}$$

Under this assumption it is proved in Ref. 9 (see also Ref. 11) that there exists a QMS with generator given in the form:

$$\mathcal{L}(x) = i[H, x] - \frac{1}{2}\sum_{i,j}\left(L_{i,j}{}^* L_{i,j} x - 2 L_{i,j}{}^* x L_{i,j} + x L_{i,j}{}^* L_{i,j}\right), \tag{22}$$

with H and $L_{i,j}$ introduced in (19), (20), $x \in \mathfrak{A}(\mathfrak{h}_0)$ and $\mathfrak{v}$ is a core domain for $\mathcal{L}(x)$.

A straightforward computation shows that

$$[H, L_{i,j}] = (E_j - E_i) L_{i,j},$$

for all i,j so that Corollary 2.1 applies. We state below the consequences of this remark.

Given $\eta \in \mathbf{S}$, $i,j \in \mathbb{N}$, define $c_{i,j}(\eta) = \eta(i)(1 - \eta(j))\gamma_{i,j}$. The following result is proved in Ref. 9.

Proposition 5.1. *For each $x \in \mathfrak{A}(\mathfrak{h}_0)$ the unbounded operator*

$$\mathcal{L}(x) = i[H, x] - \frac{1}{2}\sum_{i,j}\left(L_{i,j}^* L_{i,j} x - 2 L_{i,j}^* x L_{i,j} + x L_{i,j}^* L_{i,j}\right), \tag{23}$$

whose domain contains the dense manifold $\mathfrak{v}$, is the generator of a quantum Feller semigroup $\mathcal{T}$ on the C^*-algebra $\mathfrak{A}(\mathfrak{h}_0)$. This semigroup is extended into a σ-weak continuous QMS defined on the whole algebra $\mathfrak{L}(\mathfrak{h})$.

Moreover, the semigroup is an adiabatic perturbation of the automorphism group generated by H. The semigroup leaves the algebra $W^*(H)$ invariant. The semigroup $\mathbf{T}$ restricted to $W^*(H)$ corresponds to that of a classical exclusion process with generator

$$Lf(\eta) = \sum_{i,j} c_{i,j}(\eta)\left(f(\eta + 1_j - 1_i) - f(\eta)\right), \tag{24}$$

for all bounded cylindrical function $f : \mathbf{S} \to \mathbb{R}$.

Acknowledgments

This research has been partially funded by FONDECYT grants 1030552, 7060196 and PBCT-ACT13. The first author is greatly indebted to Ana Bela Cruzeiro, Jorge Rezende and Jean-Claude Zambrini, organizers of the meeting *Stochastic Analysis in Mathematical Physics*, for their warm hospitality in Lisbon.

References

1. L. Accardi, Y. G. Lu, I. Volovich. *Quantum theory and its stochastic limit.* Springer Verlag, 2002.
2. O. Bratteli, D. W. Robinson. *Operator algebras and quantum statistical mechanics. 2*, second ed., Texts and Monographs in Physics, Springer-Verlag, Berlin, 1997, Equilibrium states. Models in quantum statistical mechanics.
3. E. Christensen, D. E. Evans. Cohomology of operator algebras and quantum dynamical semigroups. *J. Lon. Math. Soc.* **20** (1979), 358–368.
4. E. B. Davies. Markovian master equations ii. *Math. Ann.* **219** (1976), 147–158.
5. F. Fagnola, R. Rebolledo. The approach to equilibrium of a class of quantum dynamical semigroups. *Inf. Dim. Anal. Q. Prob. and Rel. Topics* **4** (1998), 1–12.
6. F. Fagnola, M. Skeide. *Restrictions of cp-semigroups to maximal commutative subalgebras.*
7. G. Lindblad. On the generators of quantum dynamical semigroups. *Comm. Math. Phys.* **48** (1976), 119–130.
8. G. K. Pedersen. *Analysis now.* Graduate Texts in Mathematics, vol. 118, Springer-Verlag, 1989.
9. R. Rebolledo. Decoherence of quantum Markov semigroups. *Ann. Inst. H. Poincaré Probab. Statist.* **41** (**3**) (2005), 349–373.
10. R. Rebolledo. *Complete positivity and the Markov structure of open quantum systems.* Open quantum systems. II, Lecture Notes in Math., vol. 1881, Springer, Berlin, 2006, 149–182.
11. D. Spehner, J. Bellissard. A kinetic model of quantum jumps. *J. Statist. Phys.* **104** (**3–4**) (2001), 525–572.
12. D. Spehner, J. Bellissard. *The quantum jumps approach for infinitely many states.* Modern challenges in quantum optics (Santiago, 2000), Lecture Notes in Phys., vol. 575, Springer, Berlin, 2001, 355–376.
13. D. Spehner, M. Orszag. Quantum jump dynamics in cavity QED. *J. Math. Phys.* **43** (**7**) (2002), 3511–3537.
14. H. Spohn. Kinetic equations from hamiltonian dynamics: Markovian limits. *Rev. Mod. Phys.* **53** (1980), 569–615.

Gauge theory in two dimensions: Topological, geometric and probabilistic aspects

Ambar N. Sengupta

Department of Mathematics, Louisiana State University,
Baton Rouge, LA 70803, USA
E-mail: sengupta@math.lsu.edu

We present a description of two dimensional Yang-Mills gauge theory on the plane and on compact surfaces, examining the topological, geometric and probabilistic aspects.

Keywords: Yang-Mills, gauge theory, QCD, large-N

1. Introduction

Two dimensional Yang-Mills theory has proved to be a surprisingly rich model, despite, or possibly because of, its simplicity and tractability both in classical and quantum forms. The purpose of this article is to give a largely self-contained introduction to classical and quantum Yang-Mills theory on the plane and on compact surfaces, along with its relationship to Chern-Simons theory, illustrating some of the directions of current and recent research activity.

2. Yang-Mills Gauge Theory

The physical concept of a gauge field is mathematically modeled by the notion of a connection on a principal bundle. In this section we shall go over a rapid account of the differential geometric notions describing a gauge field (for a full account, see, for instance, Bleecker[10]).

Consider a smooth manifold M, to be thought of as spacetime. Let G be a Lie group, viewed as the group of symmetries of a particle field. The latter may be thought of, locally, as a function on M with values in a vector space E on which there is a representation ρ of G; a change of 'gauge' (analogous to a change of coordinates) alters ψ by multiplication by $\rho(g)$, where g is a 'local' gauge transformation, i.e. a function on M with values in G. To deal with such fields in a unified way, it is best to introduce a principal G-bundle over M. This is a manifold P, with a smooth surjection

$$\pi : P \to M$$

and a smooth right action of G:

$$P \times G \to P : (p, g) \mapsto R_g p = pg,$$

such that π is locally trivial, i.e. each point of M has a neighborhood U for which there is a diffeomeorphism

$$\phi : U \times G \to \pi^{-1}(U)$$

satisfying

$$\pi\phi(u, g) = u, \qquad \phi(u, gh) = \phi(u, g)h, \qquad \text{for all } u \in U, \text{ and } g, h \in G.$$

It will be convenient for later use to note here that a bundle over M is often specified by an indexing set I (which may have a structure, rather than just be an abstract set), an open covering $\{U_\alpha\}_{\alpha \in I}$ of M, and for each $\alpha, \beta \in I$ for which $U_\alpha \cap U_\beta \neq \emptyset$, a diffeomorphism

$$\phi_{\alpha\beta} : U_\alpha \cap U_\beta \to G$$

such that

$$\phi_{\alpha\beta}(x)\phi_{\beta\gamma}(x) = \phi_{\alpha\gamma}(x) \quad \text{for all } x \in U_\alpha \cap U_\beta \cap U_\gamma. \tag{1}$$

The principal bundle P may be recovered or constructed from this data by taking the set $\cup_{\alpha \in I}\{\alpha\} \times U_\alpha \times G$ and identifying (α, x, g) with (β, y, h) if $x = y \in U_\alpha \cap U_\beta$ and $\phi_{\beta\alpha}(x)g = h$. The elements of P are the equivalence classes $[\alpha, x, g]$, and the map $\pi : P \to M : [\alpha, x, g] \to x$ is the bundle projection and $[\alpha, x, g]k = [\alpha, x, gk]$ specifies the right G-action on P. The map $\phi_\alpha : U_\alpha \times G \to P : (x, g) \mapsto [\alpha, x, g]$ is a local trivialization. This construction is traditional (see, for instance, Steenrod[53]), but lends itself to an interesting application in the context of Yang-Mills as we shall see later.

A particle field is then described by a function $\psi : P \to E$, where E is as before, satisfying the equivariance property

$$\psi(pg) = \rho(g^{-1})\psi(p) \tag{2}$$

which is physically interpreted as the gauge transformation behavior of the field ψ. In terms of local trivializations, the value of the field over a point $x \in M$ would be described by an equivalence class $[\alpha, x, v]$ with $v = \psi([\alpha, x, e]) \in E$, and (α, x, v) declared equivalent to (β, y, w) if $y = x \in$

$U_\alpha \cap U_\beta$ and $w = \rho(\phi_{\beta\alpha}(x))v$. An equivalent point of view is to consider the space E_P of all equivalence classes $[p, v] \in P \times E$, with $[p, v] = [pg, \rho(g)^{-1}v]$ for all $(g, p, v) \in G \times P \times E$, which is a vector bundle $E_P \to M : [p, v] \mapsto \pi(p)$, and then ψ corresponds to the section of this bundle given by $M \to E_P : x \mapsto [p, \psi(p)]$ for any $p \in \pi^{-1}(x)$. These traditional considerations will be useful in an unorthodox context later in defining the Chern-Simons action (45).

The interaction of the particle field with a gauge 'force' field is described through a Lagrangian, which involves derivatives of ψ. This derivative is a 'covariant derivative', with a behavior under gauge transformations controlled through a field ω which is the gauge field. Mathematically, ω is a *connection* on P, i.e. a smooth 1-form on P with values in the Lie algebra LG of G such that

$$R_g^* \omega = \mathrm{Ad}(g^{-1})\omega \quad \text{and} \quad \omega(pH) = H \qquad (3)$$

for all $p \in P$ and $H \in LG$, where

$$pH = \frac{d}{dt}\bigg|_{t=0} p \cdot \exp(tH).$$

The tangent space $T_p P$ splits into the *vertical subspace* $\ker d\pi_p$ and the horizontal subspace $\ker \omega_p$:

$$T_p P = V_p \oplus H_p^\omega, \quad \text{where } V_p = \ker d\pi_p \text{ and } H_p^\omega = \ker \omega_p. \qquad (4)$$

A path in P is said to be horizontal, or *parallel*, with respect to ω, if its tangent vector is horizontal at every point. Thus, if

$$c : [a, b] \to M$$

is a piecewise smooth path, and u a point on the initial fiber $\pi^{-1}(c(a))$, then there is a unique piecewise smooth path $\tilde{c}_u : [a, b] \to P$ such that

$$\pi \circ \tilde{c}_u = c, \qquad \tilde{c}_u(a) = u,$$

and $\tilde{c}_u$ is composed of horizontal pieces. The path $\tilde{c}_u$ is called the *horizontal lift* of c through u, and $\tilde{c}_u(t)$ is the *parallel transport* of u along c up to time t.

The point $\tilde{c}_u(b)$ lies over the end point $c(b)$. If c is a loop then there is a unique element h in G for which

$$\tilde{c}_u(b) = \tilde{c}_u(a) h.$$

This h is the *holonomy* of ω around the loop c, beginning at u, and we denote this

$$h_u(c; \omega).$$

The property (3) implies that

$$h_{ug}(c;\omega) = g^{-1}h_u(c;\omega)g. \tag{5}$$

In all cases of interest, the gauge group G is a matrix group, and we have then the *Wilson loop observable*

$$\mathrm{Tr}\big(h(c;\omega)\big) \tag{6}$$

where we have dropped the initial point u as it does not affect the value of the trace of the holonomy.

If ρ is a representation of G on a vector space E, then there is induced in the obvious way an 'action' of the Lie algebra LG on E, and this allows us to multiply, or 'wedge', E-valued forms and LG-valued forms. If η is an E-valued k-form on P then the *covariant derivative* $D^\omega \eta$ is the E-valued $(k+1)$-form on P given by

$$D^\omega \eta = d\eta + \omega \wedge \eta \tag{7}$$

The holonomy around a small loop is, roughly, the integral of the *curvature* of ω over the region enclosed by the loop. More technically, the curvature Ω^ω is the LG-valued 2-form given by

$$\Omega^\omega = D^\omega \omega = d\omega + \frac{1}{2}[\omega \wedge \omega] \tag{8}$$

(see discussion following (26) below for explanation of notation). This LG-valued 2-form is 0 when evaluated on a pair of vectors at least one of which is vertical, and is equivariant:

$$R_g^* \Omega^\omega = \mathrm{Ad}(g^{-1})\Omega^\omega. \tag{9}$$

The connection ω is said to be *flat* if its curvature is 0, and in this case holonomies around null-homotopic loops are the identity.

Now consider an Ad-invariant metric $\langle \cdot, \cdot \rangle$ on LG. This, along with a metric on M, induces a metric $\tilde{g}$ on the bundle P in the natural way. Then we can form the curvature 'squared':

$$\langle \Omega^\omega, \Omega^\omega \rangle = \tilde{g}(\Omega^\omega, \Omega^\omega)$$

which, by equivariance of the curvature form and the Ad-invariance of the metric on LG, descends to a well-defined function on the base manifold M. The Yang-Mills action functional is

$$S_{\mathrm{YM}}(\omega) = \frac{1}{2g^2} \int_M \langle \Omega^\omega, \Omega^\omega \rangle \, d\mathrm{vol} \tag{10}$$

where the integration is with respect to the volume induced by the metric on M. The parameter g is a physical quantity which we will refer to as the coupling constant. The Yang-Mills equations are the variational equations for this action.

A *gauge transformation* is a diffeomorphism

$$\phi : P \to P$$

which preserves each fiber and is G-equivariant:

$$\pi \circ \phi = \phi \quad \text{and} \quad R_g \circ \phi = \phi \circ R_g \quad \text{for all } g \in G.$$

The gauge transformation ϕ is specified uniquely through the function $g_\phi : P \to G$, given by

$$\phi(p) = pg_\phi(p), \quad \text{for all } p \in P, \tag{11}$$

which satisfies the equivariance condition

$$g_\phi(ph) = h^{-1} g_\phi(p) h, \quad \text{for all } p \in P \text{ and } h \in G.$$

Conversely, if a smooth function $g : P \to G$ satisfies $g(ph) = h^{-1} g(p) h$ for all $p \in P$ and $h \in G$, then the map

$$\phi_g : p \mapsto pg(p) \tag{12}$$

is a gauge transformation. The group of all gauge transformations is usually denoted $\mathcal{G}$; note that the group law of composition corresponds to pointwise multiplication: $g_{\phi \circ \tau} = g_\phi g_\tau$. If $o \in M$ is a basepoint on M, it is often convenient to consider $\mathcal{G}_o$, the subgroup of $\mathcal{G}$, which acts as identity on $\pi^{-1}(o)$. The group $\mathcal{G}$ acts on the infinite dimensional affine space $\mathcal{A}$ of all connections by pullbacks:

$$\mathcal{A} \times \mathcal{G} \to \mathcal{A} : (\omega, \phi) \mapsto \phi^* \omega = \omega^{g_\phi} \stackrel{\text{def}}{=} \operatorname{Ad}(g_\phi^{-1}) \omega + g_\phi^{-1} dg_\phi \tag{13}$$

The Yang-Mills action functional and physical observables such as the Wilson loop observables are all gauge invariant.

Gauge groups of interest in physics are products of groups such as $U(N)$ and $SU(N)$, for $N \in \{1, 2, 3\}$. In the case of the electromagnetic field, $G = U(1)$ and the connection form is $\omega = i\frac{e}{\hbar} A$, where A is the electromagnetic potential, $\hbar$ is Planck's constant divided by 2π, and e is the charge of the particle (electron) to which the field is coupled. The curvature Ω^ω descends to an ordinary 2-form on spacetime, and corresponds to $\frac{e}{\hbar}$ times the electromagnetic field strength form F.

Moving from the classical theory of the gauge field to the quantum theory leads to the consideration of functional integrals of the form

$$\int_{\mathcal{A}} f(\omega) e^{-S_{\text{YM}}(\omega)} D\omega,$$

where f is a gauge invariant function such as the product of traces, in various representations, of holonomies around loops. The integral can be viewed as being over the quotient space $\mathcal{A}/\mathcal{G}$. Here the base manifold M is now a Riemannian manifold rather than Lorentzian (for the latter, the functional integrals are Feynman functional integrals, having an i in the exponent). More specifically, one would like to compute, or at least gain an understanding of, the averages:

$$W(C_1,...C_k) = \frac{1}{Z_g} \int_{\mathcal{A}/\mathcal{G}} \prod_{j=1}^{k} \text{Tr}\left(h(C_j;\omega)\right) e^{-S_{\text{YM}}(\omega)}[D\omega], \qquad (14)$$

with $[D\omega]$ denoting the formal 'Lebesgue measure' on $\mathcal{A}$ pushed down to $\mathcal{A}/\mathcal{G}$. Here the traces may be in different representations of the group G. The formal probability measure μ_g on $\mathcal{A}/\mathcal{G}$, or on $\mathcal{A}/\mathcal{G}_o$, given through

$$d\mu_g([\omega]) = \frac{1}{Z_g} e^{-\frac{1}{2g^2}\|\Omega^\omega\|_{L^2}^2}[D\omega], \qquad (15)$$

is usually called the *Yang-Mills measure*.

These integrals can be computed exactly when $\dim M = 2$, as we will describe in the following section, and the Yang-Mills measure then has a rigorous definition.

3. Wilson loop integrals in two dimensions

The Yang-Mills action is, on the face of it, quartic in the connection form ω. However, when we pass to the quotient $\mathcal{A}/\mathcal{G}$, a simplification results when the base manifold M is two dimensional. This is most convincingly demonstrated in the case $M = \mathbb{R}^2$. In this case, for any connection ω we can choose, for instance, *radial gauge*, a section

$$s_\omega : \mathbb{R}^2 \to P$$

(a smooth map with $\pi \circ s_\omega(x) = x$ for all points $x \in \mathbb{R}^2$) which maps each radial ray from the origin o into an ω-horizontal curve in P emanating from a chosen initial point $u \in \pi^{-1}(o)$. Then let F^ω be the LG-valued function on $\mathbb{R}^2$ specified by

$$\omega \mapsto s_\omega^* \Omega^\omega = F^\omega d\sigma, \qquad (16)$$

where σ is the area 2-form on $\mathbb{R}^2$. Then

$$\omega \mapsto F^\omega$$

identifies $\mathcal{A}/\mathcal{G}_o$ with the linear space of smooth LG-valued functions on $\mathbb{R}^2$ and the Yang-Mills measure becomes the well-defined Gaussian measure on the space of functions F given by

$$d\mu_g(F) = \frac{1}{Z_g} e^{-\frac{1}{2g^2}\|F\|_{L^2}^2} DF \qquad (17)$$

This measure lives on a completion of the Hilbert space of LG-valued L^2 functions on the plane, and the corresponding connections are therefore quite 'rough'. In particular, the differential equation defining parallel transport needs to be reinterpreted as a stochastic differential equation. The holonomy $h(C;\omega)$ (basepoint fixed at u once and for all) is then a G-valued random variable. The Wilson loop expectation values work out explicitly using two facts:

- If C is a piecewise smooth simple closed loop in the plane C then the holonomy $h(C)$ is a G-valued random variable with distribution $Q_{g^2 S}(x)dx$, where S is the area enclosed by C, and $Q_t(x)$ is the solution of the heat equation

$$\frac{\partial Q_t(x)}{\partial t} = \frac{1}{2}\Delta Q_t(x), \qquad \lim_{t\downarrow 0} \int_G f(x) Q_t(x)\, dx = f(e),$$

for all continuous functions f on G, with dx being unit mass Haar measure on G, and Δ is the Laplacian operator on G with respect to the chosen invariant inner product.

- If C_1 and C_2 are simple loops enclosing disjoint planar regions then $h(C_1)$ and $h(C_2)$ are independent random variables.

In the simplest case, for a simple closed loop C in the plane,

$$\int f(h(C))\, d\mu_g = \int_G f(x) Q_{g^2 S}(x)\, dx \qquad (18)$$

with S denoting the area enclosed by C. In particular, for the group $G = U(N)$,

$$W_N(C) = e^{-Ng^2 S/2} \qquad (19)$$

where

$$W_N(C) = \int \frac{1}{N} \mathrm{Tr}(h(C))\, d\mu_g.$$

Now consider the case where $M = \Sigma$, a closed oriented surface with Riemannian structure. We will follow Lévy's development[38] of the discrete Yang-Mills measure. Let $\pi : \tilde{G} \to G : \tilde{x} \mapsto x$ be the universal covering of G. Let $\mathbb{G}$ be a triangulation of Σ, or a graph, with $\mathbb{V}$ the set of vertices, $\mathbb{E}$ the set of (oriented) edges, and $\mathbb{F}$ the set of faces. We assume that each face is diffeomorphic to the unit disk, and the boundary of each face is a simple loop in the graph. Following Lévy,[38] define a discrete connection over $\mathbb{G}$ to be a map $\tilde{h} : \mathbb{E} \to \tilde{G}$ satisfying

$$\pi\big(\tilde{h}(e^{-1})\big) = \pi\big(\tilde{h}(e)\big)^{-1} \quad \text{for every edge } e \in \mathbb{E} \tag{20}$$

where e^{-1} denotes the edge e with reversed orientation. One should interpret $\tilde{h}(e)$ as the parallel transport along edge e of a continuum connection lifted to $\tilde{G}$ appropriately. Let

$$\mathcal{A}_\mathbb{G}$$

be the set of all such connections over $\mathbb{G}$. Note that this is naturally a subset of $\tilde{G}^\mathbb{E}$, and indeed can be viewed as $\tilde{G}^{\mathbb{E}_+}$, where $\mathbb{E}_+$ is the set of edges each counted only once with a particular chosen orientation; in particular, we have a unit mass Haar product measure on $\mathcal{A}_\mathbb{G}$. Define the *discrete Yang-Mills measure* μ_g^{YM} for the graph $\mathbb{G}$, by requiring that for any continuous function f on $\mathcal{A}_\mathbb{G}$, we have

$$\int_{\mathcal{A}_\mathbb{G}} f\, d\mu_g^{\mathrm{YM}} = \frac{1}{Z_g} \int f(h) \prod_{F \in \mathbb{F}} Q_{g^2|F|}\big(\tilde{h}(\partial F)\big)\, dh, \tag{21}$$

where $|F|$ is the area of the face F according to the Riemannian metric on Σ, and Z_g a normalizing constant to ensure that $\mu_g^{\mathrm{YM}}(\mathcal{A}_\mathbb{G})$ is 1. This is the discrete Yang-Mills measure for connections over all principal $\tilde{G}$-bundles over Σ. However, when G is not simply connected there are different topological classes of bundles, each specified through an element $z \in \ker(\tilde{G} \to G)$. For such z, again following Lévy,[38]

$$\mathcal{A}_\mathbb{G}^z = \left\{ \tilde{h} \in \mathcal{A}_\mathbb{G} : \prod_{e \in \mathbb{E}_+} \tilde{h}(e)\tilde{h}(e^{-1}) = z \right\} \tag{22}$$

corresponds to the set of connections on the principal G-bundle over Σ classified topologically by z. The Yang-Mills measure $\mu_{z,g}^{\mathrm{YM}}$ on $\mathcal{A}_\mathbb{G}^z$ is then simply

$$d\mu_{z,g}^{\mathrm{YM}}(\tilde{h}) = c_z 1_{\mathcal{A}_\mathbb{G}^z}(\tilde{h}) d\mu_g^{\mathrm{YM}}(\tilde{h}), \tag{23}$$

where c_z is again chosen to normalize the measure to have total mass 1. A key feature of the discrete Yang-Mills measure is that it is unaltered by subdivision of faces (plaquettes), which is why we do not need to index μ_g^{YM} by the graph $\mathbb{G}$; this invariance was observed by Migdal[42] in the physics literature. Lévy[36,38] constructed a continuum measure from these discrete measures and showed that the continuum measure thus constructed agrees with that constructed in Ref. 46. The continuum construction of the Yang-Mills measure relies on earlier work by Driver[19] and others;[25] a separate approach to the continuum Yang-Mills functional integral in two dimensions was developed by Fine[20,21] (see also Ashtekar et al.[5]).

The normalizing factor which appears in the loop expectation values is given, for a simply connected group G and a closed oriented surface of genus γ, by

$$\int_{G^{2\gamma}} Q_{g^2 S}\big(K_\gamma(x)\big)\,dx \qquad (24)$$

where K_γ is the product commutator function

$$K_\gamma(a_1, b_1, ..., a_\gamma, b_\gamma) = b_\gamma^{-1} a_\gamma^{-1} b_\gamma a_\gamma ... b_1^{-1} b_1^{-1} b_1 a_1 \qquad (25)$$

which plays the role of 'total curvature' of a discrete connection whose holonomies around 2γ standard generators of $\pi_1(\Sigma)$ are given by $a_1, b_1, ..., a_\gamma, b_\gamma$.

4. Yang-Mills on surfaces and Chern-Simons: the symplectic limit

In this section we will describe how Yang-Mills theory on surfaces fits into a hierarchy of topological/geometric field theories in low dimensions. For a detailed development of Chern-Simons theory from the point of view of topological field theory we refer to Freed[22] from which we borrow many ideas, and some notation, here. Most of our discussion below applies to trivial principal bundles (see Ref. 23 for non-trivial bundles). In Albeverio et al.,[4] the relationship between the Chern-Simons and Yang-Mills systems was explored using the method of exterior differential systems of Griffiths[24] in the calculus of variations.

One of our purposes here is to also verify that the 'correct' (from the Chern-Simons point of view) inner-product on the Lie algebra of the gauge group $SU(N)$ to use for two-dimensional Yang-Mills is independent of N. This is a small but significant fact when considering the large N limit of the Yang-Mills theory.

4.1. From four dimensions to three: the Chern-Simons form

Let $P_W \to W$ be a principal G-bundle over a manifold W. Then for any connection ω on P_W we have the curvature 2-form Ω^ω which gives rise to an $LG \otimes LG$-valued 4-form by wedging

$$\Omega^\omega \wedge \Omega^\omega$$

Now consider a metric $\langle \cdot, \cdot \rangle$ on LG which is Ad-invariant. This produces a 4-form

$$\langle \Omega^\omega \wedge \Omega^\omega \rangle$$

which, by Ad-invariance, descends to a 4-form on W which we denote again by $\langle \Omega^\omega \wedge \Omega^\omega \rangle$. The latter, a Chern-Weil form, is a closed 4-form and specifies a cohomology class in $H^4(W)$ determined by the bundle $P_W \to W$ (independent of the choice of ω).

The *Chern-Simons* 3-form $cs(\omega)$ on the bundle space P_W is given by

$$cs(\omega) = \left\langle \omega \wedge d\omega + \frac{1}{3}\omega \wedge [\omega \wedge \omega] \right\rangle = \left\langle \omega \wedge \Omega^\omega - \frac{1}{6}\omega \wedge [\omega \wedge \omega] \right\rangle \qquad (26)$$

Here wedge products of LG-valued forms, and expressions such as $[\omega \wedge \omega]$, may be computed by expressing the forms in terms of a basis of LG and ordinary differential forms. For example, writing ω as $\sum_a \omega^a E_a$, where $\{E_a\}$ is a basis of LG, the 2-form $[\omega \wedge \omega]$, whose value on a pair of vectors (X,Y) is $2[\omega(X), \omega(Y)]$, is $\sum_{a,b} \omega^a \wedge \omega^b [E_a, E_b]$. If LG is realized as a Lie algebra of matrices, then $[\omega \wedge \omega]$ works out to be $2\omega \wedge \omega$, this being computed using matrix multiplication.

The fundamental property Ref. 13 of the Chern-Simons form is that its exterior differential is the closed 4-form $\langle \Omega^\omega \wedge \Omega^\omega \rangle$ on the bundle space:

$$dcs(\omega) = \langle \Omega^\omega \wedge \Omega^\omega \rangle \qquad (27)$$

Unlike the Chern-Weil form, $cs(\omega)$ does not descend naturally to a form on W, i.e. if $s : W \to P_W$ is a section then $s^*cs(\omega)$ depends on s. If $g : P \to G$ specifies a gauge transformation $p \mapsto pg(p)$ then a lengthy but straightforward computation shows that

$$cs(\omega^g) - cs(\omega) = d\langle \omega \wedge (dg)g^{-1} \rangle - \frac{1}{6}\langle g^{-1}dg \wedge [g^{-1}dg \wedge g^{-1}dg] \rangle \qquad (28)$$

If we split a closed oriented 4-manifold W into two 4-manifolds W_1 and W_2, glued along a compact oriented 3-manifold Y, and if P_W admits sections s_1 over W_1 and s over W_2, then

$$\int_W \langle \Omega^\omega \wedge \Omega^\omega \rangle = \int_Y \left(s_1^* cs(\omega) - s^* cs(\omega) \right) \tag{29}$$

Now the sections s_1 and s are related by a gauge transformation g specified through a smooth map

$$\tilde{g}: Y \to G \tag{30}$$

in the sense that (notation as in (11) and (12))

$$s_1(y) = s(y)\tilde{g}(y) = \phi_g(s(y)), \quad \text{for all } y \in Y. \tag{31}$$

Then, after using Stokes' theorem, the term on the right in (29) works out to

$$-\frac{1}{6} \int_Y \tilde{g}^* \sigma \tag{32}$$

where σ is the 3-form on G given by

$$\sigma = \langle h^{-1} dh \wedge [h^{-1} dh \wedge h^{-1} dh] \rangle, \tag{33}$$

with $h: G \to G$ being the identity map. By choosing the metric on LG appropriately, we can ensure that this quantity is always an integer times (a convenient normalizing factor) $8\pi^2$. For example, if $G = SU(2)$, and the inner-product on LG given by

$$\langle H, K \rangle = -\text{Tr}(HK), \tag{34}$$

computation of the volume of $SU(2)$ shows that

$$\int_{SU(2)} \sigma = -48\pi^2 \tag{35}$$

(The sign on the right just fixes an orientation for $SU(2)$.) This computation can be worked out conveniently through the 2-to-1 parametrization of $SU(2)$ given by $h = k_\phi a_\theta k_\psi$, with $(\phi, \theta, \psi) \in (0, 2\pi) \times (0, \pi) \times (0, 2\pi)$, where

$$k_t = \begin{pmatrix} e^{it} & 0 \\ 0 & e^{-it} \end{pmatrix}$$

and

$$a_\theta = \begin{pmatrix} \cos\frac{\theta}{2} & i \sin\frac{\theta}{2} \\ i \sin\frac{\theta}{2} & \cos\frac{\theta}{2} \end{pmatrix}$$

Putting all this together we see that

$$\int_W \left[\frac{1}{8\pi^2} s^* cs(\omega^g) - \frac{1}{8\pi^2} s^* cs(\omega) \right] = -\int_Y \frac{1}{48\pi^2} \tilde{g}^* \sigma \in \mathbb{Z} \tag{36}$$

More generally, we assume that the metric $\langle \cdot, \cdot \rangle$ is such that

$$\frac{1}{8\pi^2} \langle \Omega^\omega \wedge \Omega^\omega \rangle$$

is an integer cohomology class for every closed oriented four-manifold W (this condition can be restated more completely in terms of the classifying space $B\tilde{G}$; see Witten[57]). For instance, for $G = SU(N)$, the properly scaled metric is (according to Witten[57]):

$$\langle H, K \rangle = -\text{Tr}(HK) \tag{37}$$

Let

$$CS(s,\omega) = \frac{1}{8\pi^2} \int_Y s^* cs(\omega) \tag{38}$$

where $s : Y \to P$ is a smooth global section (assumed to exist); the discussions above show that when s is altered, $CS(\omega)$ is changed by an integer. Thus, for any integer $k \in \mathbb{Z}$, the quantity

$$e^{2\pi k i CS(s,\omega)} \in U(1) \tag{39}$$

is independent of the section s, and thus gauge invariant.

4.2. From three dimensions to two: the $U(1)$ bundle over the space of connections on a surface

Now consider a compact oriented 3-manifold Y with boundary X, a closed oriented 2-manifold. We follow Freed's approach.[22] We assume that G is connected, compact, and simply connected; a consequence is that a principal G-bundle over any manifold of dimension ≤ 3 is necessarily trivial. For any smooth sections $s_1, s : Y \to P$, with $s_1 = s\tilde{g}$, we have on using (28) and notation explained therein,

$$\int_Y s^* cs(\omega^g) - \int_Y s^* cs(\omega) = \int_X \langle s^*\omega \wedge (d\tilde{g})\tilde{g}^{-1} \rangle - \int_Y \frac{1}{6} \tilde{g}^* \sigma \tag{40}$$

Let

$$CS(s,\omega) = \frac{1}{8\pi^2} \int_Y s^* cs(\omega) \tag{41}$$

Then, for any integer k,

$$e^{2\pi k i CS(sg,\omega)} = e^{2\pi k i CS(s,\omega)} \phi_{s\tilde{g},s}(\omega) \tag{42}$$

where

$$\phi_{s\tilde{g},s}(\omega) = e^{2\pi k i \left[\frac{1}{8\pi^2} \int_X \langle s^*\omega \wedge (d\tilde{g})\tilde{g}^{-1} \rangle - \int_Y \frac{1}{48\pi^2} \tilde{g}^*\sigma\right]} \tag{43}$$

The second term in the exponent on the right is determined, due to integrality of σ, by $\tilde{g}|X$, and is independent of the extension of $\tilde{g}$ to Y. Thus $\phi_{s\tilde{g},s}(\omega)$ is determined by s, ω, and $\tilde{g}$ on the two-manifold X.

These data specify a principal $U(1)$ bundle over the space $\mathcal{A}_X$ of connections on the bundle $P_X \to X$ (restriction of P over X), as follows. Let I be the set of all smooth sections $s : X \to P_X$. Taking this as indexing set, if $s_1, s \in I$ then, denoting by $\tilde{g} : X \to G$ the function for which $s_1 = s\tilde{g}$, we define $\phi_{s_1,s}$ as above. Thus, $\phi_{s_1,s}(\omega)$ is given by

$$\phi_{s_1,s}(\omega) = e^{2\pi kiCS(s_1,\omega)} e^{-2\pi kiCS(s,\omega)}, \tag{44}$$

where, on the right, the Chern-Simons actions are computed for extensions of $s^*\omega$ and $s^*\omega^g$ over a 3-manifold Y whose boundary is X. If a different 3-manifold Y' is chosen then the value of $\phi_{s_1,s}(\omega)$ remains the same, because it gets multiplied by

$$e^{2\pi kiCS_{Y'\cup -Y}(s_1,\omega)} e^{-2\pi kiCS_{Y'\cup -Y}(s,\omega)},$$

with obvious notation, and we have seen that this is 1. The expression (44) makes it clear that $\{\phi_{s_1,s}\}_{s,s_1 \in I}$ satisfies the cocycle condition (1) and thus *specifies a principal $U(1)$-bundle over the space $\mathcal{A}_X$ of connections on $P_X \to X$.*

Note that the integrality condition on k (which goes back to the integrality property of the inner-product on LG) is what leads to the $U(1)$ bundle.

The principal $U(1)$-bundle constructed along with the natural representation of $U(1)$ on $\mathbb{C}$, yields a line bundle $\mathbb{L}$ over $\mathcal{A}_X$, as described more generally in the context of (2). If Y is a 3-manifold with boundary X then for any connection ω on the bundle over Y, we have a well-defined element

$$e^{2\pi kiCS(\omega)} \stackrel{\text{def}}{=} [s, e^{2\pi kiCS(s,\omega)}] \tag{45}$$

in the $U(1)$-bundle over $\mathcal{A}_X$ in the fiber over $\omega|X$. In this way (following Freed[22]), the *exponentiated Chern-Simons action over an oriented 3-manifold Y with boundary X appears as a section of the line bundle $\mathbb{L}$ over $\mathcal{A}_X$.*

4.3. Connection on the $U(1)$ bundle over the space of connections

The method of geometric quantization also requires a connection on the $U(1)$-bundle (over phase space). The connection is here generated again using the Chern-Simons action. Let

$$[0,1] \to \mathcal{A}_X : t \mapsto \omega_t$$

be a path of connections, such that $(t,p) \mapsto \omega_t(p)$ is smooth. Then this specifies a connection ω on the bundle

$$[0,1] \times P \to [0,1] \times X$$

in the obvious way (parallel transport in the t direction is trivial). We define parallel transport along the path $t \mapsto \omega_t$ over $\mathcal{A}_X$ geometrically as follows: consider any 3-manifold Y with boundary X, and a principal G-bundle $P_Y \to Y$ with connection $\omega_{0,Y}$ which restricts to the given bundle over X and ω_0, and similarly consider $\omega_{1,Y}$; then parallel-transporting $e^{2\pi k i CS(\omega_{0,Y})}$ along the path will yield

$$e^{2\pi k i CS(\omega_{1,Y})} e^{2\pi k i CS(\tilde{\omega})}$$

where $\tilde{\omega}$ is the connection over $(Y) \cup (X \times [0,1]) \cup (-Y)$, glued along X, obtained by combining ω, $\omega_{0,Y}$ and $\omega_{1,Y}$. In terms of a trivialization of the bundle specified through a section s of P over X, parallel transport is given by multiplication by

$$e^{2\pi k i CS(\tilde{s}, \tilde{\omega})} \tag{46}$$

where $\tilde{s}$ is the induced trivialization of $[0,1] \times P \to [0,1] \times X$. Observe that (indicating by the subscript X the differential over X)

$$d\tilde{s}^*\tilde{\omega} = d_X s^*\omega_t + dt \wedge \frac{\partial s^*\omega_t}{\partial t}.$$

A simple computation then shows

$$CS(\tilde{s},\tilde{\omega}) = -\frac{1}{8\pi^2} \int_{[0,1]} \left(\int_X \left\langle \omega_t \wedge \frac{\partial \omega_t}{\partial t} \right\rangle \right) \wedge dt. \tag{47}$$

Viewing the Lie algebra of $U(1)$ as $i\mathbb{R}$, the parallel transport for a $U(1)$ connection along a path is e^{-P}, where P is the integral of the connection form along the path, we see that the connection form θ on the $U(1)$ bundle over $\mathcal{A}_X$ is given explicitly by

$$\theta|_\omega(A) = 2\pi i \frac{k}{8\pi^2} \int_X \langle \omega \wedge A \rangle, \tag{48}$$

for any connection $\omega \in \mathcal{A}_X$ and any vector A tangent to $\mathcal{A}_X$ at ω (such an A is simply an LG-valued 1-form on P which vanishes on vertical vectors and satisfies $R_g^* A = \text{Ad}(g^{-1})A$ for every $g \in G$. The curvature of this is given by the $i\mathbb{R}$-valued 2-form $\Theta = d\theta$ specified explicitly on $\mathcal{A}_X$ by

$$\Theta(A, B) = A(\Theta(B)) - B(\Theta(A))$$

(where A and B are treated as 'constant' vector fields on the affine space $\mathcal{A}_X$). This yields

$$\Theta(A, B) = 2\pi i \frac{k}{4\pi^2} \int_X \langle A \wedge B \rangle \tag{49}$$

for all $A, B \in T_\omega \mathcal{A}_X$. In keeping with the Bohr-Sommerfeld quantization conditions, we should consider the the symplectic form

$$\frac{1}{2\pi i}\Theta = \frac{k}{4\pi^2} \int_X \langle A \wedge B \rangle \tag{50}$$

This is precisely, with correct scaling factors, the symplectic structure used by Witten [equation (2.29) in Ref. 57] with $k = 1$.

In the context of geometric quantization it is more common to consider the Hermitian line bundle associated to the principal G-bundle over $\mathcal{A}_X$ constructed here, and view the connection as a connection on this line bundle. From this point of view one might as well simply consider the case $k = 1$, since the case of general $k \in \mathbb{Z}$ arises from different representations of $U(1)$, i.e. are tensor powers of the $k = 1$ line bundle (and its conjugate).

4.4. From Chern-Simons to Yang-Mills on a surface

The original gauge invariance of $e^{2\pi k i CS(\omega)}$ transfers to an easily-checked gauge invariance of the symplectic structure Θ on the space of connections. Thus, we have the group $\mathcal{G}$ of all gauge transformations acting symplectically on the affine space $\mathcal{A}_X$. As is well known, this action has a moment map:

$$J : \mathcal{A}_X \to (L\mathcal{G})^* : \omega \mapsto \frac{k}{4\pi^2}\Omega^\omega \tag{51}$$

where we have identified the dual of the infinite dimensional Lie algebra $L\mathcal{G}$ with the space of LG-valued Ad-equivariant functions on the bundle space P. This fact is readily checked using Stokes' theorem:

$$\langle J'(\omega)A, H \rangle = \frac{k}{4\pi^2} \int_X \langle (dA + [\omega \wedge A]), H \rangle = \frac{1}{2\pi i}\Theta(A, dH + [\omega, H])$$

The *Yang-Mills* action now can be seen as the norm-squared of the moment map:

$$S_{\text{YM}}(\omega) = \frac{1}{2g^2} \left\| \frac{4\pi^2}{k} J \right\|^2 \tag{52}$$

where $\|J\|^2$ is computed as an L^2-norm squared.

We have been discussing Chern-Simons theory in terms of its action, i.e. the integral of the Lagrangian. The Hamiltonian picture works with the *phase space*, i.e. the space of extrema of the Chern-Simons action. A fairly straightforward computation shows that the extrema are flat connections. If we consider the 3-manifold

$$Y = [0, T] \times \Sigma,$$

where Σ is a closed oriented surface, then the phase space, after quotienting out the gauge symmetries, may be identified as the moduli space of flat connections over Σ, which in turn is $J^{-1}(0)/\mathcal{G}$. It is a stratified space, with maximal stratum $\mathcal{M}^0$ which is a symplectic manifold with symplectic structure induced by $\frac{1}{2\pi i}\Theta$. We denote this symplectic structure by $\overline{\Omega}$ when k is set to 1, i.e. it is induced by the symplectic structure on $\mathcal{A}_X$ given by

$$\frac{1}{4\pi^2} \int_X \langle A \wedge B \rangle \tag{53}$$

4.5. The symplectic limit

The formal Chern-Simons path integral

$$\int_{\mathcal{A}_Y} e^{2\pi k i CS(\omega)} D\omega$$

is naturally of interest in the quantization of Chern-Simons theory (for progress on a rigorous meaning for Chern-Simons functional integrals see Hahn[29,30]). The path integral may be analyzed in the $k \to \infty$ limit by means of its behavior at the extremal of CS, i.e. on the moduli space of flat connections. This is also what results when we examine the limit of the Yang-Mills measure

$$\frac{1}{N_g} e^{-\frac{1}{2g^2}\|\Omega^\omega\|^2} D\omega,$$

(with N_g a formal normalizing factor) for connections over the surface X, in the limit $g \to 0$.

Formally, it is clear that the limiting measure, if it is meaningful, should live on those connections where Ω^ω is 0, i.e. the flat connections. Quotienting by gauge transformations yields the moduli space $\mathcal{M}^0$ of flat connections. For a compact oriented surface Σ of genus $\gamma \geq 1$, the fundamental group $\pi_1(\Sigma, o)$, where o is any chosen basepoint, is generated by the homotopy classes of loops $A_1, B_1, ..., A_\gamma, B_\gamma$ subject to the constraint that the word

$B_\gamma^{-1}A_\gamma^{-1}B_\gamma A_\gamma \ldots B_1^{-1}A_1^{-1}B_1 A_1$ is the identity in homotopy. Considering a (compact, connected,) simply connected gauge group G (so that a principal G-bundle over Σ is necessarily trivial), each flat connection is specified, up to gauge transformations, by the holonomies around the loops A_i, B_i. In this way, $\mathcal{M}^0$ is then identified with the subset of $G^{2\gamma}$, modulo conjugation by G, consisting of all $(a_1, b_1, \ldots, a_\gamma, b_\gamma)$ satisfying

$$b_\gamma^{-1}a_\gamma^{-1}b_\gamma a_\gamma \ldots b_1^{-1}a_1^{-1}b_1 a_1 = e.$$

Recalling our description of the Yang-Mills measure in terms of the heat kernel Q_t on G, we have the following result:[49]

Theorem 4.1. *Consider a closed, oriented Riemannian two-manifold of genus $\gamma \geq 2$, and assume that G is a compact, connected, simply-connected Lie group, with Lie algebra equipped with an Ad-invariant metric. Let f be a G-conjugation invariant continuous function on $G^{2\gamma}$, and $\tilde{f}$ the induced function on subsets of $G^{2\gamma}/G$. Then*

$$\lim_{t \downarrow 0} \int_{G^{2\gamma}} f(x) Q_t(K_\gamma(x))\, dx = \frac{(2\pi)^n}{|Z(G)| \mathrm{vol}(G)]^{2\gamma - 2}} \int_{\mathcal{M}^0} \tilde{f}\, d\mathrm{vol}_{\overline{\Omega}}, \quad (54)$$

where $|Z(G)|$ is the number of elements in the center $Z(G)$ of G, and $\mathrm{vol}_{\overline{\Omega}}$ is the symplectic volume form $\frac{1}{n!}\overline{\Omega}^n$ on the space $\mathcal{M}^0$ whose dimension is $2n = (2\gamma - 2)\dim G$.

With $f = 1$ this yields Witten's volume formula (formula (4.72) in Ref. 58)

$$\mathrm{vol}_{\overline{\Omega}}(\mathcal{M}^0) = \frac{|Z(G)| \mathrm{vol}(G)]^{2\gamma - 2}}{(2\pi)^n} \sum_\alpha \frac{1}{(\dim \alpha)^{2\gamma - 2}} \quad (55)$$

where the sum is over all non-isomorphic irreducible representations α of G. Specialized to $G = SU(2)$, this gives the symplectic volume of the moduli space of flat $SU(2)$ connections over a closed genus γ surface to be the rational number $\frac{2^{\gamma-1}}{(2\gamma-2)!}(-1)^\gamma B_{2\gamma-2}$, where B_k is the k-th Bernoulli number. (Note that keeping track of all the factors of 2π pays off in reaching this rational number!)

5. Concluding Remarks

We have given an overview of the geometric and topological aspects of two-dimensional Yang-Mills theory and described how they relate to the Yang-Mills probability measure.

Many physical systems involving a parameter N have asymptotic limiting forms as $N \to \infty$, even though such a limit may not have a clear physical meaning. For the case of Yang-Mills gauge theory with gauge group $U(N)$, the limit as $N \to \infty$ (holding $g^2 N$ fixed, where g is the coupling constant) has been of particular interest since the path breaking work of 't Hooft.[54] We refer to the recent review[50] for more details on the large N limit of Yang-Mills in two dimensions. A key observation is that letting $N \to \infty$, while holding $\tilde{g}^2 = g^2 N$ fixed, yields meaningful finite limits of all Wilson loop expectation values. There is also good reason to believe (see Singer[52]) that a meaningful $N = \infty$ theory also exists, possibly with relevance to Yang-Mills gauge theory in higher dimensions as well. Free probability theory (see, for instance, Voiculescu et al.[55] and Biane[9]) is likely to play a significant role here.

The partition function for $U(N)$ gauge theory on a genus γ surface is the normalizing constant we have come across:

$$Z_\gamma = \sum_\alpha (\dim \alpha)^{2-2\gamma} e^{-\tilde{g}^2 S c_2(\alpha)/(2N)}$$

where the sum is over all distinct irreducible representations α of $U(N)$, which may be viewed as a sum over the corresponding Young tablueaux (which parametrize the irreducble representations), and $c_2(\alpha)$ is the quadratic Casimir for α. This sum may be viewed naturally as a statistical mechanical partition function for a system whose states are given by the Young tableaux. This point of view leads to the study of Schur-Weyl duality for $U(N)$ gauge theory (see, for example, Ref. 1) and to the study of phase transitions in the parameter $\tilde{g}^2 S$ as $N \uparrow \infty$, viewed as a thermodynamic limit.

The references below present a sample of relevant works, and does not aspire to be a comprehensive bibliography.

Acknowledgments

The author thanks the anonymous referee for useful comments. Research support from the U.S. National Science Foundation (Grant DMS-0601141) is gratefully acknowledged.

References

1. A. D'Adda, P. Provero. Two-Dimensional Gauge Theories of the Symmetric Group S_n in the Large-n Limit. *Comm. Math. Phys.* **245** (2004), 1–25.
2. O. Aharony, S. S. Gubser, J. Maldacena, H. Ooguri, Y. Oz. *Large N Field Theories, String Theory and Gravity.* http://arxiv.org/abs/hep-th/9905111
3. S. Albeverio, H. Holden, R. Hoegh-Krohn. Markov cosurfaces and gauge fields. *Acta. Phys. Austr.*, [Supl.] XXVI (1984), 211–231.

4. S. Albeverio, A. Hahn, A. N. Sengupta. Rigorous Feynman Path Integrals, with Applications to Quantum Theory, Gauge Fields, and Topological Invariants, in *Stochastic Analysis and Mathematical Physics* (Editors: R. Rebolledo, J. Rezende, J.-C. Zambrini), World Scientific (2004), 1–60.
5. A. Ashtekar, J. Lewandowski, D. Marolf, J. Mourão, T. Thiemann. $SU(N)$ quantum Yang-Mills theory in two dimensions: a complete solution. *J. Math. Phys.* **38** (1997), 5453–5482.
6. M. F. Atiyah. *The Geometry and Physics of Knots.* Cambridge University Press (1990).
7. M. F. Atiyah, R. Bott. The Yang-Mills Equations over Riemann Surfaces. *Phil. Trans. R. Soc. Lond. A* **308** (1982), 523–615.
8. J. Baez, W. Taylor. *Strings and Two-Dimensional QCD for finite N*, hep-th/941041.
9. Ph. Biane. Free Bownian Motion, Free Stochastic Calculus and Random Matrices. *Fields Institute Communications* **12** (1997).
10. D. Bleecker. *Gauge Theory and Variational Principles.* Addison-Wesley Pub Co (1981).
11. N. Bralić. Exact Computation of Loop Averages in Two-dimensional Yang-Mills Theory. *Phys. Rev. D* **22 (12)**, (1980), 3090–3103.
12. S. Cordes, G. Moore, S. Ramgoolam. *Lectures on 2-d Yang-Mills Theory, Equivariant Cohomology, and Topological Field Theories*, hep-th/9411210.
13. S.-S. Chern and J. Simons. Characteristic forms and geometric invariants. *Annals of Math.* **99** (1974), 48–69.
14. S. de Haro, S. Ramgoolam, A. Torrielli. *Large N Expansion of q-Deformed Two-Dimensional Yang-Mills Theory and Hecke Algebras*, hep-th/0603056.
15. P. Deligne, P. Etingof, D. S. Freed, et al (Editors). *Quantum Fields and Strings: A Course for Mathematicians. Vols I and II.* American Math. Soc. and Institute for Advanced Study (1999).
16. P. A. M. Dirac. Quantised Singularities in the Eletcromagnetic Field. *Proc. Roy. Soc. Lond.* **A 133** (1931), 60–72.
17. M. R. Douglas. Large N Quantum Field Theory and Matrix Models. In *Free Probability Theory*, D.-V. Voiculescu (Ed.), Amer. Math. Soc. (1997).
18. M. R. Douglas, V. A. Kazakov. Large N phase transition in continuum QCD_2. *Phys. Lett. B* **319** (1993), 219–230.
19. B. K. Driver. YM_2: Continuum Expectations, Lattice Convergence, and Lassos. *Comm. Math. Phys.* **123** (1989), 575–616.
20. D. Fine. Quantum Yang-Mills on the Two-Sphere. *Comm. Math. Phys.* **134** (1990), 273–292.
21. D. Fine. Quantum Yang-Mills on a Riemann Surfaces. *Comm. Math. Phys.* **140** (1991), 321–338.
22. D. Freed. Classical Chern-Simons Theory, Part I, *Adv. Math.* **113** (1995) 237–303.
23. D. Freed. *Classical Chern-Simons Theory, Part II.* Available at http://www.ma.utexas.edu/users/dafr/.
24. P. A. Griffiths. *Exterior Differential Systems and the Calculus of Variations.* Birkhäuser Boston (1983).

25. L. Gross, C. King, A. Sengupta. Two Dimensional Yang-Mills Theory via Stochastic Differential Equations. *Ann. Phys. (N.Y.)* **194** (1989), 65–112.
26. D. Gross. Two-dimensional QCD as a String Theory. *Nucl. Phys. B* **400** (1993), 161–180.
27. D. Gross, W. I. Taylor. Two-dimensional QCD is a String Theory. *Nucl. Phys. B* **400** (1993), 181–210.
28. D. Gross, W. Taylor. Twists and Wilson loops in the string theory of two-dimensional QCD. *Nucl. Phys. B* **403** (1993), 395–452.
29. A. Hahn. Chern-Simons models on $S^2 \times S^1$, torus gauge fixing, and link invariants I. *J. Geom. Phys.* **53 (3)** (2005), 275–314.
30. A. Hahn. The Wilson loop observables of Chern-Simons theory on $\mathbb{R}^3$ in axial gauge. *Comm. Math. Phys.* **248 (3)** (2004), 467–499.
31. V. A. Kazakov. Wilson Loop Average for an Arbitrary Contour in Two-dimensional $U(N)$ Gauge Theory. *Nucl. Phys. B* **179** (1981), 283–292.
32. V. A. Kazakov, I. K. Kostov. Nonlinear Strings in Two-dimensional $U(\infty)$ Gauge Theory. *Nucl. Phys. B* **176** (1980), 199–215.
33. V. A. Kazakov, I. K. Kostov. Computation of the Wilson Loop Functional in Two-dimensional $U(\infty)$ Lattice Gauge Theory. *Phys. Lett. B* **105 (6)** (1981), 453–456.
34. T. P. Killingback. Quantum Yang-Mills Theory on Riemann Surfaces and Conformal Field Theory. *Physics Letters B* **223** (1989), 357–364.
35. S. Klimek, W. Kondracki. A Construction of Two-Dimensional Quantum Chromodynamics. *Comm. Math. Phys.* **113** (1987), 389–402.
36. T. Lévy. The Yang-Mills measure for compact surfaces. *Memoirs Amer. Math. Soc.* **166 (790)** (2003).
37. T. Lévy. Wilson loops in the light of spin networks. *Journal of Geometry and Physics* **52** (2004), 382–397. http://arxiv.org/abs/math-ph/0306059
38. T. Lévy. Discrete and continuous Yang-Mills measure for non-trivial bundles over compact surfaces, *Probability Theory and Related Fields* **136 (2)** (2006), 171-202.
39. T. Lévy, J. Norris. Large Deviations for the Yang-Mills Measure on a Compact Surface, *Comm. Math. Phys.* **261 (2)** (2006), 405-450.
40. Y. Makeenko, A. Migdal. Quantum chromodynamics as dynamics of loops. *Nucl. Phys. B* **188** (1981), 269–316.
41. M. L. Mehta. *Random Matrices*, Elsevier (2004).
42. A. A. Migdal. Recursion Equations in Gauge Field Theories. *Sov. Phys. JETP* **42** (1975), 413–418.
43. A. A. Migdal. Properties of the Loop Average in QCD. *Annals of Physics* **126** (1980), 279–290.
44. A. A. Migdal. Exact Equivalence of Multicolor QCD to the Fermi String Theory. *Physics Letters B* **96** (1980), 333–336.
45. G. Paffuti, P. Rossi. A Solution of Wilson's Loop Equation in Lattice QCD_2. *Physics Letters B* **92** (1980), 321-323.
46. A. Sengupta. Quantum Gauge Theory on Compact Surfaces. *Ann. Phys. (NY)* **221** (1993), 17-52.
47. A. Sengupta. Gauge Theory on Compact Surfaces. *Memoirs of the Amer.*

Math. Soc. **126 (600)** (1997).
48. A. Sengupta. Yang-Mills on Surfaces with Boundary: Quantum Theory and Symplectic Limit, *Comm. Math. Phys.* **183** (1997), 661–706.
49. A. N. Sengupta. The volume measure of flat connections as limit of the Yang-Mills measure. *J. Geom. Phys.* **47** (2003), 398–426.
50. A. N. Sengupta. Traces in Two-Dimensional QCD: The Large-N Limit. In *Traces in gemeotry, number theory and quantum fields*, Eds. S. Albeverio, M. Marcolli, S. Paycha, to be published by Vieweg.
51. A. N. Sengupta. A functional integral applied to topology and algebra. In *XIVth International Congress on Mathematical Physics: Lisbon 28 July - 2 August 2003*, World Scientific Publishing Company (2006).
52. I. M. Singer. On the Master Field in Two Dimensions. In *Functional Analysis on the Eve of the 21st Century, Volume I*, Ed. S. Gindikin et al., Birkhäuser (1995).
53. N. Steenrod. *The Topology of Fiber Bundles*. Princeton University Press (1999).
54. G. 't Hooft. A planar diagram theory for strong interactions. *Nucl. Phys. B* **72** (1974), 461–473.
55. D. V. Voiculescu, K. J. Dykema and A. Nica. *Free Random Variables*. CRM Monograph Series **1**, Amer. Math. Soc. (1992).
56. K. Wilson. Confinement of Quarks. *Phys. Rev. D* **10** (1974), 2445.
57. E. Witten. On Quantum Gauge Theories in Two Dimensions. *Comm. Math. Phys.* **141** (1991), 153–209.
58. E. Witten. Two Dimensional Quantum Gauge Theory revisited. *J. Geom. Phys.* **9** (1992), 303–368.
59. T. T. Wu, C. N. Yang. Concept of non-integrable phase factors and global formulation of gauge fields. *Phys. Rev. D* **12** (1975), 3845–3857.
60. F. Xu. A Random Matrix Model from Two-Dimensional Yang-Mills Theory. *Comm. Math. Phys.* **190** (1997), 287–307.

Near extinction of solution caused by strong absorption on a fine-grained set

V. V. Yurinsky

Universidade da Beira Interior,
6201-001 Covilhã, Portugal
E-mail: yurinsky@ubi.pt

A. L. Piatnitski

HiN, Narvik University College,
P.O. Box 385, N-8505 Narvik, Norway
E-mail: andrey@sci.lebedev.ru

This article considers the large-time behaviour of solutions of a nonlinear parabolic equation modelling heat transfer in a medium with a highly oscillating absorption coefficient (e.g., one that is generated by a "random chessboard" or periodic). The minimal size of a cube where absorption is substantial irrespective of its position is the small parameter of the problem.

If the absorption coefficient is separated from zero on a disperse fine-grained set, the behaviour of a solution is shown to imitate extinction in finite time — even when the lack of absorption on a massive set makes *bona fide* extinction impossible. Namely, as long as the instantaneous value of thermal energy exceeds a small threshold value, it admits a decreasing majorant that vanishes after a finite time. For energies below this threshold, this majorant becomes unapplicable; it can be replaced by one which decays fast, but remains strictly positive.

It is also shown that the Dirichlet problem for a quasi-linear heat equation with nonlinear absorption term can be homogenized.

Keywords: Absorption-diffusion equation, medium with microstructure, random chessboard, finite time extinction, homogenization

1. Introduction

This paper is dedicated to the large-time behaviour of solutions of a model boundary problem describing heat transfer (or diffusion) in a bounded domain $G \Subset \mathbb{R}^d$ with regular boundary: for $t > 0$ and $x \in G$

$$\partial_t \left(|u_\varepsilon|^{\gamma-2} u_\varepsilon \right) = \mathrm{div} \left(a \left| \nabla u_\varepsilon \right|^{p-2} \nabla u_\varepsilon \right) - S_\varepsilon^\sigma \left| u_\varepsilon \right|^{\sigma-2} u_\varepsilon, \qquad (1.1)$$

where the matrix $a(x,t,u)$ is symmetric, bounded and strictly positive definite:

$$0 < A_* |\xi|^2 \leq a_\varepsilon(x,t,u)\xi \cdot \xi \leq A^* |\xi|^2. \quad (1.2)$$

The initial and boundary conditions are

$$u_\varepsilon|_{t=0} = u_0 \in L^{\gamma+k}(G), \ k \in \mathbb{R}_+, \ u_\varepsilon|_{\partial G} = 0. \quad (1.3)$$

The exponents of nonlinearities are constant and satisfy the condition

$$1 < \sigma < \gamma \leq p \leq d. \quad (1.4)$$

The absorption coefficient $S_\varepsilon(x) \geq 0$ does not depend on time t and is a highly oscillating function of the spatial variable (e.g., ε-periodic or random). The existence of a solution is assumed as a prerequisite (some pertinent existence and uniqueness theorems are briefly discussed in §2.1).

Equation (1.1) is obviously a modification of the classical heat equation. In the context of thermal transport, the nonlinearities model the dependence of properties of the heat carrier on its temperature. For instance, the choice of $\gamma = p = 2$ and $\sigma < 2$ in Eq. (1.4) corresponds to a material which combines constant thermal conductivity with heat absorption that increases as it cools down. Other admissible choices of exponents maintain a similar property of the underlying physical process.

The influence of nonlinear absorption can produce a qualitative difference in the behaviour of solution. Without external forcing, the linear heat equation describes the process of cooling which never ends completely. By contrast, Eq. (1.1) with exponents of Eq. (1.4) defines a solution which vanishes completely after a finite time if the absorption coefficient S_ε is separated from zero on all domain (see, e.g., Ref. 1 (Ch.2 §2.3), the survey Ref. 2 and Refs. 3,4).

However, the above qualitative difference in behaviour of solutions is sensitive to seemingly slight alterations of the problem. Extinction of the solution in finite time does not occur if the non-absorbing set $F_\varepsilon = \{S_\varepsilon = 0\}$ has positive measure.

For instance, a solution of the simplest quasi-linear heat equation ($\gamma = p = 2$, $a = Id$) cannot vanish on all domain in finite time if its initial value is positive on an arbitrarily small ball contained in F_ε. This follows, e.g., from the Feynman-Kac representation for the solution of the heat equation because a Brownian trajectory can protract its stay in the ball indefinitely (albeit with a very low probability of late exit).

Theorem 2.1 of Sec. 2, which is the main result of this paper, shows that the behaviour of a solution to Eq. (1.1) can, nevertheless, imitate extinction in finite time even when the set F_ε has positive measure. This can occur under the following condition.

Condition 1.1 (Dispersiveness). *There exists a function $K(\varepsilon) \in \mathbb{N}$ such that $\lim_{\varepsilon \to 0} \varepsilon K(\varepsilon) = 0$ and*

$$\forall z \in \mathbb{G}(\varepsilon, K) \quad |C_{\varepsilon, K(\varepsilon), z}|^{-1} |C_{\varepsilon, K(\varepsilon), z} \cap \{S_\varepsilon > \beta\}| \geq \tau > 0, \ \beta > 0, \quad (1.5)$$

where $\mathbb{G}(\varepsilon, K) = \{z \in \mathbb{Z}^d : |C_{\varepsilon, K, z} \cap G| > 0\}$ *and*

$$C_{\varepsilon, K, z} = \{x : x - \varepsilon K z \in]0, \varepsilon K]^d\}, \ K \in \mathbb{N}, K \geq K(\varepsilon). \quad (1.6)$$

The sets $C_{\varepsilon,K,z}$ are later called εK-*blocks*. Condition 1.1 is satisfied with $K = 1$ if S_ε is ε-periodic and not identically zero. When restrictions of S_ε to individual cells $Y_{\varepsilon,z} \equiv C_{\varepsilon,1,z}$ are independent, Eq. (1.5) holds with very high probability for blocks of size $\varepsilon \ln(1/\varepsilon)$ or greater (see Appendix A.3).

The proof of Theorem 2.1 combines the energy method[1] with techniques used to establish deterministic large-volume asymptotic behaviour of the principal eigenvalue (PE) for elliptic operators with random non-negative potential.[5,6] Its approach is related to that of Refs. 3,4 which detects finite-time extinction of solutions of nonlinear parabolic equations through the study of PE's of pertinent Schrödinger operators.

It seems appropriate to show the tools used to prove Theorem 2.1 in a heuristic argument. For the linear problem $\partial_t v = \Delta v - S_\varepsilon^\sigma v$, $v|_{\partial G} = 0$, the solution's norm $\|v(t)\|_2$ decays exponentially, and its half-life is inversely proportional to the Dirichlet PE $\lambda_\varepsilon = \inf_\phi \|\phi\|_2^{-2}(\|\nabla \phi\|_2 + \|S_\varepsilon^\sigma \phi^2\|_1)$. Decay is fast if S_ε satisfies some form of Cond. 1.1 and ε is small. (For example, calculations[5,6] done for $G = [0,1]^d$ and absorption restricted to ε-balls surrounding points of a Poisson cloud with intensity $\mu \varepsilon^{-d}$ show that $\lambda_\varepsilon \asymp \varepsilon^{-2}(\ln \frac{1}{\varepsilon})^{-2/d}$ as $\varepsilon \to 0$.)

For a solution of the non-linear equation $\partial_t u = \Delta u - |u|^{\sigma-2} S^\sigma u$ under the same boundary and initial conditions, the instantaneous rate of decay is proportional to $\Lambda_\varepsilon(t) = \inf_\phi \|\phi\|_2^{-2}(\|\nabla \phi\|_2^2 + \||u|^{\sigma-2} S_\varepsilon^\sigma \phi^2\|_1)$, so the decay of norm $\|u\|_2$ should accelerate as it nears zero. If the non-absorbing set $\{S_\varepsilon = 0\}$ is empty, this results in finite time extinction.[1,3,4] Otherwise, the decay slows down when the solution becomes negligible outside $\{S_\varepsilon = 0\}$.

Theorem 2.1 provides some majorants for an appropriate functional $U(t)$ of the solution (see §2.2). It shows that the evolution of $U(t)$ includes a phase of "attempted extinction" if $\varepsilon K(\varepsilon)$ of Eq. (1.5) is small. Namely, there exists a time interval where the solution's norm admits a majorant that vanishes after a finite time.

When $U(t)$ drops below a small ε-dependent threshold value, the majorant suggestive of extinction in finite time has to be substituted by one that never vanishes. However, the decay of the solution's norm remains fast (exponential if $\gamma = p$ or as a negative power of time otherwise). The times necessary to halve its value are of order $\mathcal{O}\left(\varepsilon^\lambda\right)$, $\lambda > 0$, so the difference between true and simulated extinction may not be easy to detect numerically.

One more result of this article is Theorem 3.1 which shows that Eq. (1.1) admits homogenization in the simple case when $\gamma = p = 2$ and the absorption term is its only nonlinearity. The homogenized problem, which can be written down explicitly, is one with finite extinction time.

2. Near Extinction of Solution

2.1. Weak solutions

Notation. For vectors $x = (x^{(i)}) \in \mathbb{R}^d$, the scalar product is $x \cdot y = \sum_{i=1}^d x^{(i)} y^{(i)}$ and $|x| = (x \cdot x)^{1/2}$. The Lebesgue measure of $A \subset \mathbb{R}^d$ is $|A|$.

The gradient of a scalar function is $\nabla \phi = \left(\partial \phi / \partial x^{(i)}\right)$, and $\partial_t \phi$ is its time derivative.

Notation for monomials similar to nonlinear terms of Eq. (1.1) is

$$u^{\Diamond \mathcal{P}} \overset{\text{abbr}}{=} |u|^{\mathcal{P}-1} u, \ u \in \mathbb{R}. \qquad (2.1)$$

Obviously, $u^{\Diamond \alpha} u^{\Diamond k} = |u|^{\alpha+k}$ and $u^{\Diamond \alpha} |u|^k = u^{\Diamond (\alpha+k)}$. For a smooth function $\partial_t |u|^k = k u^{\Diamond k-1} \partial_t u$ and $\nabla |u|^k = k u^{\Diamond k-1} \nabla u$ if $\kappa - 1 \geq 0$ or $u \neq 0$.

When misunderstanding is unlikely, notation is abbreviated: $\phi(t)$ may refer to a function $\phi(x,t)$ considered as a function-valued mapping $t \mapsto \phi(x,t)$, $\int \phi$ can substitute $\int_A \phi(x)\,dx$ if the nature of the argument and the domain of integration are clear from the context. The L^p-norm of a function is always $\|\phi\|_p$. Notation of Sobolev spaces is standard.[7,8]

Definition of weak solution. A measurable function $u = u_\varepsilon(x,t)$ on $Q_{T+} = G \times]0, T_+[$ is a weak solution of Eq. (1.1) with initial and boundary conditions of Eq. (1.3) if for each test function $\zeta \in C^\infty([0,T_+]; C_0^\infty(G))$

$$\int_G u^{\Diamond(\gamma-1)}(x, T_+)\zeta(x, T_+)dx - \int_G u_0^{\Diamond(\gamma-1)}(x)\zeta(x,0)dx$$
$$= \int_{Q(T_+)} \left(u^{\Diamond(\gamma-1)} \partial_t \zeta - |\nabla u|^{p-2} a_\varepsilon \nabla u \cdot \nabla \zeta - S_\varepsilon^\sigma u^{\Diamond(\sigma-1)} \zeta\right) dxdt. \qquad (2.2)$$

The weak solutions u_ε considered below are, up to notation of exponents, those of Ref. 1 (see [Ch.2, §2.1, Def. 2.1]). Namely, the function $u^{\Diamond(1+k/p)}$

belongs to the space $V(Q_{T_+})$ for a given value of parameter $k \geq 0$, i.e.,

$$u^{\Diamond(1+k/p)} \in L^p\left(0, T_+; W_0^{1,p}(G)\right),$$

$$u \in L^\infty\left(0, T_+; L^{\gamma+k}(G)\right), \quad u \in L^{\sigma+k}(Q_{T_+}). \quad (2.3)$$

The possibility to choose the parameter $k \geq 0$ depending on the exponents γ, p, and σ proves important in the arguments below.

Some existence and uniqueness theorems. For the case of $\gamma = 2$, the methods of demonstration of existence and uniqueness theorems are classic.[7-10] For the linear equation with $\gamma = p = \sigma = 2$ and a not depending on the solution, the construction of the solution by the Galerkin method and proof of its uniqueness can be found in Ch.7 of Ref. 8.

For $\gamma = 2$ and $p \geq 2$, the existence and uniqueness of solution for Eq. (1.1) follow from known theorems[9,10] on parabolic equations containing monotone operators; the existence of a unique solution for Eq. (1.1) under the assumptions of Sec. 1 is established in Ch. 2 of Ref. 9 (Theorem 1.1 for $p > 2$, Theorem 1.4 and Examples 1.7.1-2 for $p > 1$).

For $\gamma < 2$ and $p > 1$, Eq. (1.1) is a special case of the doubly nonlinear parabolic equation $\partial_t b(u) = \nabla \cdot A(u, \nabla u) + f$, where $b : \mathbb{R} \to \mathbb{R}$ is nondecreasing and continuous (see survey Ref. 2). For $b(u) = u^{\Diamond(\gamma-a)}$, $1 < \gamma < 2$, the existence of solution follows from the results of Refs. 11,12. If for $v \in \mathbb{R}$, $\xi \in \mathbb{R}^d$ both $|A(v,\xi)| \leq c\left(|\xi|^{p-1} + |v|^{1/\gamma-1} + 1\right)$ and $(A(v,\xi) - A(v,\eta)) \cdot (\xi - \eta) \geq c|\xi - \eta|^p$, then there exists a solution with finite energy $\int_0^T \|u(t)\|_p^p dt + \sup_{t \in [0,T]} \|u^{(t)}\|_\gamma^\gamma$. It is unique[12,13] provided that $|A(b(u_1,\xi) - A(b(u_2,\xi)| \leq c|u_1 - u_2|^{1-1/p}\left(|\xi|^{p-1} + |u_1| + |u_2| + 1\right)$.

The energy inequality. The calculations of this article are based on the so-called energy inequality for weak solutions satisfying conditions Eq. (2.3) (see Lemma 2.1 in Ch.2 of Ref. 1 and Lemma 3.1 of Ref. 14):

$$\|u(S)\|_{\gamma+k}^{\gamma+k} - \|u(T)\|_{\gamma+k}^{\gamma+k} \quad (2.4)$$

$$\geq C \int_S^T \left(A_* \|\nabla u^{\Diamond(1+k/p)}(t)\|_p^p + \|S_\varepsilon u^{\Diamond 1+k/\sigma}(t)\|_\sigma^\sigma dt\right).$$

The following proposition is a special case of Lemma 3.1 in Ref. 14, adapted to integrals of $|u|^{\gamma+k}$ with $k > 0$ (its proof is included in Appendix A.1 to facilitate reading).

Lemma 2.1. *Consider a weak solution $u \in V(T_+)$ that corresponds to the initial value $u_0 \in L^{\gamma+k}(G)$. If $\limsup_{t \to 0+} \|u(t)\|_\gamma < \infty$ and $u(t)$ converges to u_0 in measure as $t \searrow 0$, then $u^{\Diamond(1+k/p)} \in V(Q_{T_+})$, inequality*

(2.4) holds true for $0 \leq S < T < T_+$, and $\|u(S)\|_{L^{\gamma+k}(G)} \leq \|u_0\|_{L^{\gamma+k}(G)}$ for $S \in]0, T_+[$.

A bound on Rayleigh quotient for absorbing cubes. The following estimate of the L^{p+k}-norm of a function over a cube partly covered by the absorbing set in terms of similarly restricted "instantaneous" diffusion and absorption terms of Eq. (2.4) is used below.

Given a time $t > 0$, an integer $L = L(t)$ is used to partition the space into "large" cubes $H_z = \{x : (L\varepsilon K)^{-1}x - z \in]0,1]^d\}$ of size $L\varepsilon K$. Each of these cubes is the union of $L(t)^d$ blocks (1.6). If condition (1.5) is satisfied and $S_\varepsilon(x) = \beta$ outside G, then $|H_z \cap \{S_\varepsilon \geq \beta\}| \geq \tau |H_z|$ for $L \geq 1$.

Lemma 2.2. *If condition (1.5) is satisfied and $|G \cap \{|\phi| > \Phi\}| \leq \frac{1}{2}\tau(L\varepsilon K)^d$ for some $\Phi > 0$, then*

$$\int_G \left(|\nabla(\phi^{\diamond 1+k/p})|^p + |S_\varepsilon \phi^{\diamond 1+k/\sigma}|^\sigma\right) \geq \frac{c_2 \|\phi\|_{p+k}^{p+k}}{\max\{(L\varepsilon K)^p, \beta^{-\sigma}\Phi^{p-\sigma}\}}, \quad (2.5)$$

where the constant c_2 is determined by p, k, τ, and d.

Remark 2.1. When S_ε is separated from zero on G, the size of cubes can be selected arbitrarily. For $\Phi < 1$ and $L\varepsilon K \leq \Phi^{1-\sigma/p}$ inequality (2.5) becomes $\|\phi^{1+k/p}\nabla\phi)\|_p^p + \|S_\varepsilon \phi^{\diamond 1+k/\sigma}\|^\sigma \geq C\Phi^{-(p-\sigma)}\|\phi\|_{p+k}^{p+k}$.
The proof of Lemma 2.2 makes use of Lemma A.2.

Proof of Lemma 2.2. By the assumption, the sets $Q_z = H_z$ and $Q_{0,z} = H_z \cap \{S_\varepsilon \geq \beta\} \cap \{|\phi| \leq \Phi\}$ satisfy the inequality $|Q_z|/|Q_{0,z}| \geq 2/\tau > 1$ for each single cube. Thus, Lemma A.2 yields the estimate

$$c_2 \int_{H_z} |\phi|^{p+k} \leq (L\varepsilon K)^p \int_{H_z} |\nabla \phi^{\diamond 1+k/p}|^p + \int_{Q_{0,z}} |\phi|^{p+k}$$

$$\leq (L\varepsilon K)^p \int_{H_z} |\nabla \phi^{\diamond 1+k/p}|^p + \int_{Q_{0,z}} \frac{S_\varepsilon^\sigma}{\beta^\sigma}\Phi^{p-\sigma}|\phi|^{k+\sigma},$$

which is equivalent to (2.5) with G replaced by any single cube H_z (being trivial if it has no common points with G). Summing these inequalities for $z \in \mathbb{Z}^d$ yields (2.5). □

2.2. The main result

Initial phase of decay. Embedding theorems for the Sobolev spaces $W_0^{1,p}(G)$ (see, e.g., §II.2 of Ref. 7) and well-known inequalities for L^p-norms imply that

$$\|\nabla v_p(t)\|_p^p \geq c\|v_p(t)\|_p^p = c\|u(t)\|_{p+k}^{p+k} \geq c\|u(t)\|_{\gamma+k}^p,$$

so Eq. (2.4) results in the well-known "differential inequality"

$$\psi(t) + \Psi \int_0^t \psi^\kappa(s)\,ds \le C \tag{2.6}$$

for $\psi(t) = \|u(t)\|_{\gamma+k}^{\gamma+k}$ with $C = \psi(0)$ and $\kappa = \frac{p+k}{\gamma+k} \ge 1$. This ensures convergence of $\psi(t)$ to zero as $t \to \infty$. (The possible forms of the majorant for $\psi(t)$ are reproduced in Lemma A.1).

Attempted extinction. Below, some functions on G are identified with their trivial extension to all space for the sake of convenience: $u_\varepsilon(x,t) = 0$, $S_\varepsilon(x) = \beta$, $\phi(x) = 0$, etc., for $x \notin G$.

Theorem 2.1. *Assume that Cond. 1.1 is satisfied. Let $u = u_\varepsilon(x,t)$ be a weak solution of boundary problem (1.1)-(1.3) with $u^{\diamond(1+k/p)} \in V(Q_{T_+})$, and consider the function*[a] $U(t) = \|u(t)\|_{\gamma+k}^{\gamma+k}$.

(a) If k and the exponents of Eq. (1.4) satisfy the inequality

$$\gamma + k > \frac{(p-\sigma)}{(\gamma-\sigma)} \frac{d}{p}(p-\gamma), \tag{2.7}$$

then for small values of εK there exists a time interval $\Delta = [t_0, t_1]$ on which the solution $u = u_\varepsilon$ of problem (1.1)-(1.3) decays at a rate characteristic of extinction in finite time: for $t \in \Delta$

$$U(t) \le U(t_0)\left(1 - (1-\kappa)\Psi(t-t_0)\right)^{1/(1-\kappa)}, \tag{2.8}$$

where $\kappa = \alpha_* \frac{p+k}{\gamma+k} + (1-\alpha_*)\frac{\sigma+k}{\gamma+k} < 1$, the number Ψ does not depend on ε, $\alpha_* = \left(1 + \frac{p}{d} \cdot \frac{\gamma+k}{p-\sigma}\right)^{-1}$, and

$$t_0 = \sup\{t > 0 : U(t) \le 1\}, \tag{2.9}$$

$$t_1 = \sup\left\{t > t_0 : U(t) > \tfrac{1}{3}\tau(\varepsilon K)^{d/\alpha_*}\right\}. \tag{2.10}$$

The length of interval Δ satisfies the inequality $T = t_1 - t_0 \le \Psi(1-\kappa)$.
(b) If $p > \gamma$, then for $t > t_1$

$$\frac{U(t_1+t)}{U(t_1)} \le \left(1 + (\tilde\kappa - 1)U^{(\tilde\kappa-1)}(t_1)\frac{t}{(\varepsilon K)^p}\right)^{-1/(\tilde\kappa-1)},$$

where the time scale characterizing decay is small for small εK:

$$\frac{U^{(\tilde\kappa-1)}(t_1)}{(\varepsilon K)^p} = 0((\varepsilon K)^{-\mathcal{P}}),\ \mathcal{P} = \frac{p}{1-\alpha_*}\left(\frac{\gamma-\sigma}{p-\sigma} - \alpha_*\right) > 0. \tag{2.11}$$

---
[a] It plays the part of a clock in the calculations to follow.

If $p = \gamma$, then $\widetilde{\kappa} = 1$ and the above majorant changes to $\exp\{-ct/(\varepsilon K)^p\}$.

2.3. Proof of Theorem 2.1

(a) The existence of a time t_0 such that $U(t) \leq 1$ for $t \geq t_0$ follows from Lemma 2.1.

We apply the Chebyshev inequality to evaluate the measure of the set where the solution is large: for each $t \geq t_0$,

$$|\{|u(x,t)| \geq \|u\|_{\gamma+k}^{1-\alpha}\}| \leq \int_G (|u(x,t)|/\|u\|_{\gamma+k}^{1-\alpha})^{\gamma+k} dx = U^\alpha(t) \leq 1. \quad (2.12)$$

The value of the free parameter $\alpha \in]0,1[$ will be specified later on.

It follows from condition (1.5) and (2.12) that

$$|H_z|^{-1}|H_z \cap \{S_\varepsilon \geq \beta\} \cap \{|u(t)| \leq \|u\|_{\gamma+k}^{1-\alpha}\}| \geq \tfrac{2}{3}\tau$$

if $t \geq t_0$ and

$$|H_z|^{1/d} = L(t)\varepsilon K \geq (3/\tau)^{1/d} U^{\alpha/d}(t). \quad (2.13)$$

As long as condition (2.13) is satisfied, Lemma 2.2 yields a minorant for the integral on the left-hand side of (2.4). Namely, this lemma is applicable to $\phi = u$ with $\Phi = B\beta^{\sigma/(p-\sigma)}\|u\|_{\gamma+k}^{1-\alpha}$, where B is one more parameter to be selected later.

Under the additional condition $(L\varepsilon K)^p \leq B^{p-\sigma}\|u(t)\|_{\gamma+k}^{(1-\alpha)(p-\sigma)}$, i.e.,

$$L(t)\varepsilon K \leq B^{1-\sigma/p} U(t)^{(1-\alpha)(1-\sigma/p)/(\gamma+k)}, \quad (2.14)$$

inequality (2.5) holds true with

$$\max\left\{(L(t)\varepsilon K)^p, \beta^{-\sigma}\Phi^{p-\sigma}\right\} = B^{p-\sigma}\|u\|_{\gamma+k}^{(p-\sigma)(1-\alpha)}. \quad (2.15)$$

Since $\|u\|_{p+k} \geq c\|u\|_{\gamma+k}$, it follows in this case that

$$\|\nabla u^{\diamond 1+k/p}\|_p^p + \|S_\varepsilon u^{\diamond 1+k/\sigma}\|_\sigma^\sigma \geq c_2 \|u\|_{p+k}^{p+k} B^{-(p-\sigma)} \|u\|_{\gamma+k}^{-(1-\alpha)(p-\sigma)}$$
$$\geq \widehat{c}(B)U^\kappa(t), \ \kappa(\alpha) = \alpha\tfrac{p+k}{\gamma+k} + (1-\alpha)\tfrac{\sigma+k}{\gamma+k} = \tfrac{\sigma+k}{\gamma+k} + \alpha\tfrac{p-\sigma}{\gamma+k} \quad (2.16)$$

When $U \leq 1$, the quantity $U^{\kappa(\alpha)}$ is a non-increasing function of α, so the right-hand side is largest for the smallest possible value of α.

For small values of $U(t)$, conditions (2.13) and (2.14) are compatible only if $\alpha/d \geq (1-\alpha)(1-\sigma/p)(\gamma+k)^{-1}$, so the best minorant corresponds to the value of α that satisfies this condition as equality:

$$\alpha_* = \left(1 + \frac{p}{d} \cdot \frac{\gamma+k}{p-\sigma}\right)^{-1}, \ p\alpha_* = d\frac{p-\sigma}{\gamma+k}(1-\alpha_*). \quad (2.17)$$

In the differential inequality (2.6) for $\psi(t) = U(t)$ that follows from (2.16), the rate of decay corresponds to extinction of $\psi(t)$:

$$\kappa(\alpha) < 1 \Leftrightarrow \alpha < (\gamma - \sigma)/(p - \sigma). \tag{2.18}$$

The exponent $\kappa(\alpha_*)$ is typical of finite time of extinction for α_* of Eq. (2.17) if the exponent $\gamma + k$ in the initial condition satisfies inequality (2.7) — in this case $\alpha_* < (\gamma - \sigma)/(p - \sigma)$ and

$$0 < 1 - \kappa(\alpha_*) = \frac{\gamma - \sigma}{\gamma + k} - \left(\frac{p - \sigma}{\gamma + k}\right)^2 \left(\frac{p}{d} + \frac{p - \sigma}{\gamma + k}\right)^{-1}.$$

An "optimal-order" minorant for the right-hand side of (2.16) results from the choice of the natural-valued function $L(t)$ as

$$L(t) = \left[(\varepsilon K)^{-1} U(t)^{\alpha_*/d}\right] = \left[(\varepsilon K)^{-1} \|u(t)\|_{\gamma+k}^{(1-\alpha_*)(1-\sigma/p)}\right] \tag{2.19}$$

where $[\,\cdot\,]$ is the integer part of a number.

As long as Eq. (2.16) is applicable with $\alpha = \alpha_*$ of (2.17), inequality (2.4) results in the differential inequality (2.6) with $\psi(t) = U(t_0 + t)$, the exponent $\kappa(\alpha_*)$ of (2.8), $C = U(t_0) = 1$, and the coefficient $\Psi = \widehat{c}(B)$.

However, estimate (2.16) applies only as long as $L(t) \geq 1$ in Eqs. (2.13) and (2.19). The phase of accelerating decay ends at time $t_1(\alpha_*)$ defined in Eq. (2.10), when

$$U(t_1) = \|u(t_1)\|_{\gamma+k}^{\gamma+k} = (\varepsilon K)^{d/\alpha_*}. \tag{2.20}$$

After that majorant (2.8) is no longer applicable.

To estimate the duration of the phase of "attempted extinction," note that it starts with $U(t_0) = 1$. Consequently, it cannot last longer than the time when the majorant vanishes (see Lemma A.1):

$$T(\alpha_*) \leq (\widehat{c}\beta^\sigma(1-\kappa))^{-1}.$$

(b) After time t_1, estimate (2.15) changes to

$$\max\left\{(\varepsilon K)^p, \beta^{-\sigma}\Phi^{p-\sigma}\right\} = (\varepsilon K)^p,$$

and (2.16) is replaced by the inequality

$$\|\nabla u^{\diamond 1+k/p}\|_p^p + \|S_\varepsilon u^{\diamond 1+k/\sigma}\|_\sigma^\sigma \geq c_2 (\varepsilon K)^{-p} \|u\|_{p+k}^{p+k}$$
$$\geq \widetilde{c}(\varepsilon K)^{-p} U^{\widetilde{\kappa}}, \ \widetilde{\kappa} = \frac{p+k}{\gamma+k} \geq 1.$$

If $\gamma < p$, then $\widetilde{\kappa} > 1$, so the corresponding differential inequality (2.6) produces the slow majorant of Eq. (2.11). By Eq. (2.20) the coefficient that accompanies t on the right-hand side of Eq. (2.11) is proportional to

$(\varepsilon K)^{-\mathcal{P}}$, and the use of formulae (2.17) and (2.18) shows that $\mathcal{P}$ is a positive number:

$$\mathcal{P} = p - (\widetilde{\kappa} - 1)d/\alpha_* = \frac{p}{1-\alpha_*}\left(\frac{\gamma-\sigma}{p-\sigma} - \alpha_*\right) > 0,$$

so the typical times for halving the majorant are small even at this phase of slower decay.

The majorant of Lemma A.1 decays exponentially if $p = \gamma$, $\widetilde{\kappa} = 1$. □

Remark 2.2. When the absorption coefficient S_ε is separated from zero, there is no need to restrict the use of the embedding inequality of Lemma 2.2 to cubes of size εK or greater. Hence (see Remark 2.1) the phase of accelerating decay continues until the solution really dies out in finite time.

3. A theorem on homogenization

In the simple case considered here, boundary problem (1.1)-(1.3) admits homogenization, which may be useful, e.g., for finding more accurate bounds on duration of the initial phases of the solution's decay.

Below $\gamma = p = 2$ and $1 < \sigma < 2$, while $a = A(x)$ does not depend on the solution or time and satisfies condition (1.2), so Eq. (1.1) reduces to the quasi-linear equation

$$\partial_t u_\varepsilon = \operatorname{div}(A\nabla u_\varepsilon) - S_\varepsilon^\sigma u_\varepsilon^{\diamond \sigma - 1} \qquad (3.1)$$

with the only nonlinearity in the absorption term (notation is that of Eq. (2.1)). The initial and boundary conditions are those of Eq. (1.3) with $k = 0$. As before, the existence of the solutions from $L^2(0,T;W_0^1(G)) \cap L^\infty(0,T;L^2(G))$ is assumed as a prerequisite.

For a function $\phi(x)$, notation $\langle\phi\rangle_{\varepsilon,K}(x)$ refers to the piecewise-constant function that equals $|C_{\varepsilon,K,z}|^{-1}\int_{C_{\varepsilon,K,z}}\phi(\xi)d\xi$ on the block $C_{\varepsilon,K,z}$ of Eq. (1.6).

Theorem 3.1. *Assume that the absorption coefficient $S_\varepsilon(x)$ is bounded, and there exists a a constant $\widehat{S}^\sigma$ such that for some $q > d$*

$$\|\langle S_{\varepsilon,K}^\sigma\rangle - \widehat{S}^\sigma\|_{L^q(G)} \leq \nu. \qquad (3.2)$$

If the function $W : G \times [0,T]$ satisfies the homogenized equation

$$\partial_t W = \operatorname{div}(A\nabla W) - \widehat{S}^\sigma |W|^{\sigma-2}W \qquad (3.3)$$

and the initial and boundary conditions of Eq. (1.2), then for small εK the difference of solutions to Eq. (3.1) and Eq. (3.3) admits the estimate

$$\|\nabla(u_\varepsilon - W)\|_{L^2(G\times[0,T])} \leq C\left(\nu\|W\|_{2d/(d-2)}^{\sigma-1} + \varepsilon K\|S^\sigma\|_\infty\|W\|_{2d/(d-2)}^{\sigma-1}\right.$$
$$\left.+(\varepsilon K/\delta)\|W\|_2^{\sigma-1} + \delta^{\sigma-1}\|\nabla W\|_2^{\sigma-1}\right),$$

where $\delta \in]0,1[$ is a free parameter.

Condition (3.2) holds true for blocks of size $K(\varepsilon) = O(\ln(1/\varepsilon))$ when S_ε is generated by a "random chessboard" structure with independent cells (see Lemma A.3). Convergence of u_ε to W follows from the inequality of the theorem if it is possible to choose $K = K(\varepsilon)$, $\delta = \delta(\varepsilon)$, and $\nu = \nu(\varepsilon)$ so that $K \to \infty$, $\nu(\varepsilon) \to 0$, $\varepsilon K/\delta \to 0$, and $\delta \to 0$ as $\varepsilon \to 0$.

Proof of Theorem 3.1. Below notation is abbreviated to $u = u_\varepsilon$, $S = S_\varepsilon$, and $\langle \phi \rangle = \langle \phi \rangle_{\varepsilon,K}$. The calculations deal with the case of $d > 2$.

The difference $V = u - W$ vanishes at $t = 0$ and satisfies the equation (in notation of Eq. (2.1))

$$\partial_t V - \nabla \cdot A\nabla V + S^\sigma(u^{\diamond\sigma-1} - W^{\diamond\sigma-1}) = \sum_{k=1}^3 H_k, \qquad (3.4)$$

where

$$H_1 = (S^\sigma - \langle S^\sigma \rangle)(W^{\diamond\sigma-1} - \langle W^{\diamond\sigma-1}\rangle),$$
$$H_2 = (S^\sigma - \langle S^\sigma\rangle)\langle W^{\diamond\sigma-1}\rangle, \quad H_3 = (\langle S^\sigma\rangle - \widehat{S}^\sigma)W^{\diamond\sigma-1}.$$

Using a sequence of smooth test functions convergent to $u - W$ in the integral identity, we arrive at the inequality

$$\int_0^T dt \int_G \nabla(u - W) \cdot A\nabla(u - W) dx$$
$$+ \int_0^T dt \int_G S^\sigma\left(u^{\diamond\sigma-1} - W^{\diamond\sigma-1}\right)(u - W)dx \leq \int_0^T I(t)dt, \quad (3.5)$$

where $I(t) = \sum_{k=1}^3 \int_G (u(x,t) - W(x,t))H_k(x,t)dx$. Both terms on the left-hand side are non-negative.

The integral containing H_3 of Eq. (3.4) is estimated using the Hölder inequality (recall that $\|\phi\|_{2d/(d-2)} \leq c\|\nabla\phi\|_2$):

$$\left|\int_G (u-W)H_3 dx\right| \leq \|u - W\|_2 \|\langle S^\sigma\rangle - \widehat{S}^\sigma\|_d \|W\|_{2d/(d-2)}^{\sigma-1}. \qquad (3.6)$$

Since for each block $\int_{C_{\varepsilon,K,z}} (S^\sigma(x,t) - \langle S^\sigma\rangle)(\langle u - W\rangle\langle W^{\diamond\sigma-1}\rangle) dx = 0$, it follows from the well-known embedding theorems that

$$\left|\int_G H_2(u - W)dx\right| = \left|\int_G (S^\sigma - \langle S^\sigma\rangle)(u - W - \langle u - W\rangle)\langle W^{\diamond\sigma-1}\rangle dx\right|$$
$$\leq c\varepsilon K \|\nabla(u - W)\|_2 \|S^\sigma - \langle S^\sigma\rangle\|_d \|W\|_{2d/(d-2)}^{\sigma-1}. \qquad (3.7)$$

Finally, we evaluate the integral containing H_1 of Eq. (3.4)

$$\int_G (u-W)H_1 dx = \int_G (u-W)(S^\sigma - \langle S^\sigma \rangle)(W^{\diamond\sigma-1} - \langle W^{\diamond\sigma-1}\rangle)dx.$$

To this end, we exploit smoothness of W and the following elementary inequality with $q = \sigma - 1 < 1$:

$$\forall a, b \in \mathbb{R} \ \left|a^{\diamond q} - b^{\diamond q}\right| \le 2 |a-b|^q, \ 0 < q \le 1. \tag{3.8}$$

The function $W^{\diamond\sigma-1}$ may not have appropriately summable gradient. It is convenient to approximate this factor in the integrand on the right hand side of (3.5) by its convolution with a smooth kernel $h \in C_0^\infty(\mathbb{R}^d)$:

$$W_\delta(x) = W_\delta(x,t) \equiv \int_{\mathbb{R}^d} W(x+\delta y) h(y) dy,$$
$$W_\delta^{\diamond\sigma-1}(x) = W_\delta^{\diamond\sigma-1}(x,t) \equiv \int_{\mathbb{R}^d} W^{\diamond\sigma-1}(x+\delta y) h(y) dy,$$

where $\delta > 0$, $h(y) \ge 0$, $\int_{\mathbb{R}^d} h(y) dy = 1$, and $h(y) = 0$ if $|y| > \eta$.
Inequality (3.8) furnishes the estimate

$$\left|W^{\diamond\sigma-1}(x+\delta y) - W^{\diamond\sigma-1}(x)\right| \le 2 |W(x+\delta y) - W(x)|^{\sigma-1},$$

so an application of the Hölder inequality to integral in y shows that t-a.e. on $[0,T]$ for $Q = 2/(\sigma-1)$

$$\|W_\delta^{\diamond\sigma-1} - W^{\diamond\sigma-1}\|_Q^Q \le c \int dx \left(\int |W(x+\delta y) - W(x)|^{\sigma-1} h(y) dy\right)^Q$$
$$\le c \int h(y) dy |W(x+\delta y) - W(x)|^2 dx.$$

It is well known (see, e.g., [15, §4.6]) that

$$\int_{\mathbb{R}^d} |W(x+\delta y) - W(x)|^2 dx \le c\delta^2 |y|^2 \|\nabla W\|_2^2,$$

so $\|W_\delta^{\diamond\sigma-1} - W^{\diamond\sigma-1}\|_Q^Q \le c\delta^2 \|\nabla W\|_2^2 \int |y|^2 h(y) dy = C\delta^2 \|\nabla W\|_2^2$ and

$$\|W_\delta^{\diamond\sigma-1} - W^{\diamond\sigma-1}\|_Q \le C\delta^{\sigma-1} \|\nabla W\|_2^{\sigma-1}. \tag{3.9}$$

The function $W_\delta^{\diamond\sigma}$ is smooth. Its gradient admits representation $\nabla W_\delta^{\diamond\sigma-1}(x) = \frac{1}{\delta} \int W^{\diamond\sigma-1}(x+\delta y) \nabla h(y) dy$, and consequently

$$\|\nabla W_\delta^{\diamond\sigma-1}\|_Q = \left(\int dx \left|\delta^{-1} \int_{\mathbb{R}^d} W^{\diamond\sigma-1}(x+\delta y) \nabla h(y) dy\right|^Q\right)^{(\sigma-1)/2}$$
$$\le c\delta^{-1} \left(\int dx \int dy |W(x+\delta y)|^2 |\nabla h(y)|^Q\right)^{(\sigma-1)/2} \le c\delta^{-1} \|W\|_2^{\sigma-1}.$$

It follows that

$$\|W_\delta^{\diamond\sigma-1} - \langle W_\delta^{\diamond\sigma-1}\rangle\|_Q \le c(\varepsilon K/\delta) \|W\|_2^{\sigma-1},$$

and Eq. (3.9) and the triangle inequality lead to the estimate

$$\|W^{\diamond\sigma-1} - \langle W^{\diamond\sigma-1}\rangle\|_Q \le C\left((\varepsilon K/\delta)\|W\|_2^{\sigma-1} + \delta^{\sigma-1}\|\nabla W\|_2^{\sigma-1}\right).$$

The inequality $\|\phi\|_{2d/(d-2)} \le c\|\nabla\phi\|_2$ leads to the conclusion that

$$\left|\int_G (u-W)H_1 dx\right| = \left|\int_G (u-W)(S^\sigma - \langle S^\sigma\rangle)(W^{\diamond\sigma-1} - \langle W^{\diamond\sigma-1}\rangle)dx\right|$$
$$\le c\|\nabla(u-W)\|_2\left(\|\langle S^\sigma\rangle - \widehat{S}^\sigma\|_d\|W\|_{2d/(d-2)}^{\sigma-1}\right.$$
$$+\varepsilon K\|S^\sigma\|_\infty \|W\|_{2d/(d-2)}^{\sigma-1}$$
$$\left.+(\varepsilon K/\delta)\|W\|_2^{\sigma-1} + \delta^{\sigma-1}\|\nabla W\|_2^{\sigma-1}\right). \qquad (3.10)$$

Equations (3.6), (3.7), and (3.10) yield the estimate of the theorem. □

Acknowlegments

This work was supported by FCT (Portugal): project POCI/MAT/55977/2004 *Integração Funcional e Aplicações* and sub-projecto *DECONT — Differential Equations of Continuum Mechanics: Solvability and Localization Effects, Numerical Methods* of Centro de Matemática UBI (2003–2005).

Appendix A.

A.1. Proof of Lemma 2.1.

(a) First, we consider the case $T_+ > T > S > 0$. We apply integral identity (2.2) to a sequence of admissible test functions for which all its integrals converge to their counterparts for the function

$$Z(x,t) = \mathcal{K}_\varepsilon * 1_{]S,T]}(t)\, \mathcal{K}_{\varepsilon,\delta}^+ * u_L^{\diamond k+1}(x,t), \qquad (A.1)$$

where $u_L = \text{sign}(u)|u| \wedge L$, the parameters $\varepsilon, \delta > 0$ are small and $L > 0$ large. As usual, $*$ denotes convolution in t. The mollifiers $\mathcal{K}_\varepsilon(t) = \frac{1}{\varepsilon}\mathcal{K}(\frac{1}{\varepsilon}t)$ and $\mathcal{K}_{\varepsilon,\delta}^+ = \mathcal{K}_\varepsilon * \frac{1}{\delta}1_{[-\delta,0]}(t)$ are smooth, and for the sake of convenience it is assumed that $\mathcal{K}(t) = \mathcal{K}(-t)$.

For small $\varepsilon, \delta > 0$, function (A.1) vanishes for $t = 0$ and $t = T_+$, so identity (2.2) results in the equality

$$\mathcal{T}_{\varepsilon,\delta} \equiv \int_{G\times[S,T]} u^{\diamond\gamma-1}\partial_t Z = \mathcal{A}_{\varepsilon,\delta} + \mathcal{B}_{\varepsilon,\delta}, \qquad (A.2)$$

where $\mathcal{A}_{\varepsilon,\delta} = \int_{G\times\mathbb{R}} \nabla Z \cdot a|u_\varepsilon|^{p-1}$ and $\mathcal{S}_{\varepsilon,\delta} = \int_{G\times\mathbb{R}} S^\sigma u^{\diamond\sigma-1} Z$.

The symmetry of $\mathcal{K}$ implies that $\int_\mathbb{R} g(s)\mathcal{K}_\varepsilon * h(s)\,ds = \int_\mathbb{R} \mathcal{K}_\varepsilon * g(t)h(t)\,dt$ and $\int_\mathbb{R} g(s)\mathcal{K}'_\varepsilon * h(s)\,ds = -\int_\mathbb{R} h(t)\mathcal{K}'_\varepsilon * g(t)\,dt$, so in (A.2)

$$\mathcal{T}_{\varepsilon,\delta} = -\int_{G\times[S,T]} \left(\mathcal{K}'_\varepsilon * u^{\diamond\gamma-1}\right)\left(\mathcal{K}^+_{\varepsilon,\delta} * u_L^{\diamond k+1}\right) dxdt,$$

$$\mathcal{A}_{\varepsilon,\delta} = \int_{G\times[S,T]} \mathcal{K}^+_{\varepsilon,\delta} * \left(|u|^k 1_{\{|u|<L\}}\nabla u\right) \cdot \mathcal{K}_\varepsilon * \left(|\nabla u|^{p-2} a\nabla u\right) dxdt,$$

$$\mathcal{B}_{\varepsilon,\delta} = \int_{G\times[S,T]} \mathcal{K}^+_{\varepsilon,\delta} * \left(u_L^{\diamond k+1}\right) \mathcal{K}_\varepsilon * \left(S_\varepsilon^\sigma u^{\diamond\sigma-1}\right) dxdt.$$

It is easily seen that

$$\lim_{\delta\searrow 0}\lim_{\varepsilon\searrow 0} \mathcal{A}_{\varepsilon,\delta} = \mathcal{A} \equiv \int_{G\times[S,T]} |u_L|^k 1_{\{|u|<L\}} \left(\nabla u \cdot |\nabla u|^{p-2} a\nabla u\right), \quad (A.3)$$

$$\lim_{\delta\searrow 0}\lim_{\varepsilon\searrow 0} \mathcal{B}_{\varepsilon,\delta} = \mathcal{B} \equiv \int_{G\times[S,T]} |u_L|^{k+1} S_\varepsilon^\sigma |u|^{\sigma-1}. \quad (A.4)$$

Integration by parts reduces the term of Eq. (A.2) containing time derivative to

$$\mathcal{T}_{\varepsilon,\delta} = \mathcal{T}_1(t)|_{t=T}^{t=S} + \mathcal{T}_2(S,T), \quad (A.5)$$

where $\mathcal{T}_1(t) = \int_G \left(\mathcal{K}_\varepsilon * u^{\diamond\gamma-1}\right)\left(\mathcal{K}^+_{\varepsilon,\delta} * u_L^{\diamond k+1}\right)$ and $\mathcal{T}_2(S,T) = \int_{G\times]S,T[} \mathcal{K}_\varepsilon * u^{\gamma-1}\partial_t \mathcal{K}^+_{\varepsilon,\delta} * u_L^{\diamond k+1}$.

For an integrable function $\lim_{\varepsilon\searrow 0} \mathcal{K}^+_{\varepsilon,\delta} * \phi(t) = \phi_\delta(t) \equiv \frac{1}{\delta}\int_t^{t+\delta} \phi(s)\,ds$ and $\lim_{\delta\searrow 0} \phi_\delta(t) = \phi(t)$ a.e. in t. Consequently, the limit of the first term in Eq. (A.5) is

$$\lim_{\delta\searrow 0}\lim_{\varepsilon\searrow 0} \mathcal{T}_1(t)|_{t=T}^{t=S} = \int_G u^{\gamma-1}(x,t) u_L^{\diamond k+1}(x,t)\,dx \,|_{t=T}^{t=S}. \quad (A.6)$$

It is easily seen that $\partial_t \mathcal{K}^+_{\varepsilon,\delta} * \phi = \frac{1}{\delta}(\mathcal{K}_\varepsilon * \phi(t+\delta) - \mathcal{K}_\varepsilon * \phi(t))$, so

$$\lim_{\varepsilon\searrow 0} \mathcal{T}_2(S,T) = \frac{1}{\delta}\int_{G\times[S,T]} u^{\gamma-1}\left(u_L^{\diamond k+1}(t+\delta) - u_L^{\diamond k+1}(t)\right) dxdt. \quad (A.7)$$

By analogy with the function $j(u)$ of Ref. 14, define for $k \geq 0$ and $u \in\,]-\infty,\infty[$ the nonnegative convex function

$$J_k(u) = \int_0^u v^{\diamond\gamma-1} d\left(v^{\diamond k+1}\right) = \frac{k+1}{k+\gamma} |u|^{\gamma+k}, \quad (A.8)$$

which satisfies the inequality (cf. Eq.(3.14) of Ref. 13)

$$D_L(t,\delta) = J_k(u_L(t+\delta)) - J_k(u_L(t)) \quad (A.9)$$
$$- \psi(u(t))\left(u_L^{\diamond k+1}(t+\delta) - u_L^{\diamond k+1}(t)\right) \geq 0.$$

Indeed, by the definition

$$D_L(t,\delta) = \int_{u_L(t)}^{u_L(t+\delta)} \left(v^{\diamond\gamma-1} - u^{\diamond\gamma-1}(t)\right)(k+1)|v|^k dv.$$

If $u_L(t+h) > u_L(t)$, then necessarily $u(t+h) > u(t)$ and $u(t) < u_L(t+h)$, so the integrand is non-negative. The integral is zero if $u_L(t+h) = u_L(t)$. If $u_L(t+h) < u_L(t)$, then $u(t+h) < u(t)$ and $u_L(t+h) < u(t)$, so the integrand is non-positive, and the integral non-negative.

It follows from Eq. (A.9) that in A.7

$$\lim_{\varepsilon \searrow 0} \mathcal{T}_2 \geq \tfrac{1}{\delta} \int_G \int_S^T (J_k(u_L(t+\delta)) - J_k(u(t)) dt) dx$$

$$= \int_G dx \left(\tfrac{1}{\delta} \int_T^{T+\delta} J_k(u_L(t)) dt - \tfrac{1}{\delta} \int_S^{S+\delta} J_k(u_L(t)) dt \right).$$

By the Lebesgue-Vitali theorem $\lim_{\delta \searrow 0} \lim_{\varepsilon \searrow 0} \mathcal{T}_2 = J_k(u_L(T)) - J_k(u_L(S))$ a.e. in S, T, so by (A.6)

$$\limsup_{\delta \searrow 0} \lim_{\varepsilon \searrow 0} \mathcal{T} \leq \int_G \left(u^{\diamond \gamma - 1}(x,t) u_L^{\diamond k+1}(x,t) - J_k(u_L(x,t)) \right) dx \Big|_{t=T}^{t=S}.$$

Combined with Eqs. (A.3) and (A.4), this latter estimate shows that for $T > S > 0$ and J_k of Eq. (A.8)

$$\int_G \left(u^{\gamma-1} u_L^{\diamond k+1}(x,t) - J_k(u_L(x,t)) \right) dx \Big|_{t=T}^{t=S} \tag{A.10}$$

$$\geq \int_{G \times [S,T]} \left(|u|^k \mathbf{1}_{\{|u|<L\}} \left(\nabla u \cdot |\nabla u|^{p-2} a \nabla u \right) + |u_L|^{k+1} S_\varepsilon^\sigma |u|^{\sigma-1} \right).$$

(b) On the left-hand side of Eq. (A.10) the integrands have the form $\mathcal{W}(u(x,s), L)$ and $\mathcal{W}(u(x,T), L)$, where for $v \in \mathbb{R}$

$$\tfrac{\gamma-1}{k+\gamma} |v_L|^{\gamma+k} \leq \mathcal{W}(v, L) = \left(|v|^{\gamma-1} - \tfrac{k+1}{k+\gamma} |v_L|^{\gamma-1} \right) |v_L|^{k+1} \leq |v|^{\gamma-1} |v_L|^{k+1}.$$

We compare now $u_0(x) = \mathcal{W}(u_0(x), L)$ with $U(x,S) = \mathcal{W}(u(x,S), L)$ for small $S > 0$.

By the assumption, $U(x,S) \to u_0(x)$ in measure as $S \searrow 0$. Moreover, $|U(x,S) - u_0(x)| \leq W$, where for a fixed $L > 0$ and an arbitrary $\mu > 0$

$$W = c_1(L, \gamma, k) \left(|u(x,S)|^{\gamma-1} + |u_0(x)|^{\gamma-1} + 1 \right) \mathbf{1}_{\{|U(S)-U_0| > \mu\}}$$

$$+ c_2(L, \gamma, k) \left(\mu^{\gamma-1} + \mu |u(x,S)|^{\gamma-1} + \mu |u_0(x)|^{\gamma-1} \right) \mathbf{1}_{\{|u(x,S)-u_0(x)| \leq \mu\}}.$$

By the Hölder inequality for each $\mu > 0$ and $\widehat{U} = \limsup_{S \searrow 0} \|u(S)\|_\gamma$

$$\limsup_{S \searrow 0} W \leq c_1 \limsup_{S \searrow 0} (\operatorname{mes} \{|U(S) - u_0| > \mu\})^{1/\gamma} \left(\widehat{U}^{\gamma-1} + \|u_0\|_\gamma^{\gamma-1} \right)$$

$$+ c_2 \left(\mu^{\gamma-1} + \mu \left(\widehat{U}^{\gamma-1} + \|u_0\|_\gamma^{\gamma-1} \right) \right),$$

which proves that for all L

$$\int_G u_0(x)\,dx = \lim_{S\to 0+} \int_G U(x,S)\,dx \geq \int_G U(x,S)\,dx.$$

If $u_0 \in L^{\gamma+k}(G)$, then $u_0(x,L) \to |u_0(x)|^{\gamma+k}$, and $|u_0(x)|^{\gamma+k}$ is a majorant for this family of functions. Hence

$$\lim_{L\to\infty} \int_G u_0(x)\,dx = \left(1 - \tfrac{k+1}{k+\gamma}\right) \int_G |u_{*,L}(x)|^{\gamma+k}\,dx.$$

By the above, for each $S \in\,]0, T_+[$

$$\int_G |u_L(x,S)|^{\gamma+k}\,dx \leq \left(1 - \tfrac{k+1}{k+\gamma}\right)^{-1} \int_G U(x,S)\,dx \leq \int_G |u_{*,L}(x)|^{\gamma+k}\,dx.$$

By the theorem on monotonic convergence $\|u(S)\|_{\gamma+k} < \infty$, and the existence of an integrable majorant justifies the passages to the limit as $L\to\infty$ in all integrals of Eq. (A.10). This proves the lemma. □

We also cite here the well-known differential inequality that plays an important part in detection of finite time extinction by the energy method (see Ref. 1, §2.1).

Lemma A.1. *If a non-increasing right-continuous function $\psi(t) \geq 0$ satisfies inequality (2.6) with $\kappa > 0$ on an interval $[0,T[$, then on this interval*

$$\psi(t) \leq F(t,\Psi,C,\kappa),$$

where

$$F(t,\Psi,C,\kappa) = \begin{cases} C\left(1 - (1-\kappa)\dfrac{\Psi t}{C^{1-\kappa}}\right)_+^{1/(1-\kappa)}, & \kappa < 1, \\ C\exp\{-\Psi t\}, & \kappa = 1, \\ C\left(1 + (\kappa-1)C^{\kappa-1}\Psi t\right)^{-1/(\kappa-1)}, & \kappa > 1. \end{cases}$$

A.2. An embedding inequality

The following embedding inequality is an adaptation to $p \neq 2$ of one well known in many forms for $p = 2$ (the proof below follows that of Ref. 6).

Lemma A.2. *Consider a bounded convex set $Q \subset \mathbb{R}^d$, $d \geq 2$, containing a subset Q_0 of positive Lebesgue measure such that $A^d = |Q|/|Q_0| \geq 1$. The following inequality holds true with $c = (2^d - 2)/(d-1)$ for each real function $u \in W^{1,p}(Q)$, $p > 1$:*

$$|u|^p_{L^p(Q)} \leq cA^{d-1}\left(\mathrm{diam}^p(Q)\,|\nabla u|^p_{L^p(Q)} + \int_{Q_0} |u|^p\right).$$

Proof. It is easily seen that

$$u(x) = |Q_0|^{-1} \int_{Q_0} u(y)dy + |Q_0|^{-1} \int_{Q_0} \int_0^1 (x-y) \cdot \nabla u\,(\xi_t)\,dtdy$$

for a smooth function u, an arbitrary pair of points $a, y \in Q$, and $\xi_t = \xi_t(x,y) = tx + (1-t)y$. Thus by Hölder's inequality

$$\|u\|_{L^p(Q)}^p \leq 2^{p-1} \left(\frac{|Q|}{|Q_0|} \int_{Q_0} \|u\|^p + \frac{D^p}{|Q_0|} \int_0^1 dt \int_Q dx \int_{Q_0} \|\nabla u\,(\xi_t)\|^p\,dy \right),$$

where $D = \mathrm{diam}\,(Q)$. For a fixed value of t, the argument ξ_t is in Q by convexity, so starting with integration in variable $\xi = \xi_t(x,y)$ (with fixed $y \in Q_0$) or $\eta = \xi_t$ (with fixed $x \in Q$) leads to the inequality

$$|Q_0|^{-1} \int_Q dx \int_{Q_0} \|\nabla u\,(\xi_t)\|^p\,dy \leq U(t)\,\|\nabla u\|_{L^p(Q)}^p,$$

where $U(t) = \min\left\{t^{-d}, (1-t)^{-d}|Q|/|Q_0|\right\}$. Integration in t yields the estimate of the lemma[b]. □

A.3. Inequalities for random chessboard

Below, we consider the simplest random model of the absorbing medium. The following lemma combines the well-known exponential inequality of S. N. Bernstein (see Ref. 16, §3.4) with the estimate $\#(\mathbb{G}) \leq c_1|G|/(\varepsilon K)^d$ for the number of pertinent blocks.

Lemma A.3. (a) *If there exist numbers $\beta, \lambda > 0$ and a family of i.i.d. binary random variables $\chi_z : \Omega \to \{0,1\}$ such that $\mathbf{P}\{\chi_z = 1\} = q > 0$ and $|\{S_\varepsilon \geq \beta\} \cap Y_{\varepsilon,z}| \geq \lambda \chi_z$ for all cells having common points with G, then there exist constants c_1 dependent on the shape of G and $c_2 = c_2(q,\lambda)$ such that for $\tau = \frac{1}{3}q\lambda$*

$$\mathbf{P}\left\{ \min_{z \in \mathbb{G}(\varepsilon,K)} \frac{|C_{\varepsilon,K,z} \cap \{S_\varepsilon < \beta\}|}{|C_{\varepsilon,K,z}|} \leq \tau \right\} \leq \frac{c_1|G|}{(\varepsilon K)^d} \exp\{-c_2 K^d\}.$$

(b) *If $S_\varepsilon(x,\omega) \in [0, S_+]$ is a bounded measurable random field and the random variables $\widehat{S}_z = |Y_{\varepsilon,z}|^{-1} \int_{Y_{\varepsilon,z}} S_\varepsilon$ are i.i.d., then in condition (3.2)*

$$\mathbf{P}\left\{ \|\langle S_\varepsilon \rangle_{\varepsilon,K} - \mathbf{E}\widehat{S}_0 \|_{L^\infty} \geq \nu \right\} \leq \frac{c_1|G|}{(\varepsilon K)^d} \exp\{-\widehat{c}\nu^2 K^d\}.$$

---

[b]With $c(d) = \max_{A \geq 1} \frac{A^{-d+1}}{d-1}\left((A+1)^d - A^d - 1\right)$, which is the constant of the lemma.

References

1. S. N. Antontsev, J. Díaz and S. Shmarev, *Energy Methods for Free Boundary Problems (Applications to Nonlinear PDES and Fluid Mechanics)*, Progress in Nonlinear Differential Equations and Their Applications, Vol. 48 (Birkhauser, Boston, 2002).
2. J. I. Díaz, *Extracta Mat.* **16**, 303 (2001).
3. Y. Belaud, *Electron. J. Differential Equations* **3**, 9 (2001), Conf. 8.
4. Y. Belaud, B. Helffer and L. Veron, *Ann. Inst. H. Poincaré Anal. Non Linéaire* **18**, 43 (2001).
5. A.-S. Sznitman, *Brownian Motion, Obstacles and Random Media* (Springer-Verlag, Berlin-Heidelberg-New York, 1998).
6. V. V. Yurinsky, *Probab. Theory Related Fields* **114**, 151 (1999).
7. O. A. Ladyzhenskaya, V. A. Solonnikov and N. N. Ural'tseva, *Linear and Quasilinear Equations of Prabolic Type*, Translations of Mathematical Monographs, Vol. 23 (AMS, Providence, 1967). (Translation from the Russian by S. Smith).
8. L. C. Evans, *Partial Differential Equations*, Graduate Studies in Mathematics, Vol. 19 (AMS, Providence, 1998).
9. J.-L. Lions, *Quelques méthodes de résolution des problèmes aux limites non lineaires* (Dunod, Paris, 1969).
10. R. E. Showalter, *Monotone Operators in Banach Spaces and Nonlinear Partial Differential Equations*, Mathematical Surveys and Monographs, Vol. 49 (AMS, Providence, 1997).
11. A. W. Alt and S. Luckhaus, *Math. Z.* **193**, 311 (1982).
12. A. V. Ivanov and J. F. Rodrigues, *Preprint* No. 69, (M. Planck-Institut Leipzig, 1998).
13. F. Otto, *J. Diff. Eq.* **131**, 20 (1996).
14. J. I. Díaz and L. Veron, *Trans. Amer. Math. Soc.* **290**, 787 (1985).
15. S. M. Nikolsky, *Approximation of Functions of Several Variables and Imbedding Theorems*, (Springer, Berlin, 1975).[c]
16. V. V. Petrov, *Sums of Independent Random Variables* (Springer-Verlag, Berlin-New York, 1975).

---

[c] 2nd edn. in Russian: (Nauka, Moscow, 1977)